When
Death
Becomes
Life

When
Death
Becomes
Life

Notes from a Transplant Surgeon

JOSHUA D. MEZRICH

HARPER

An Imprint of HarperCollins*Publishers*

HarperCollins books may be purchased for educational, business, or sales promotional use. For information, please email the Special Markets Department at SPsales@harpercollins.com.

FIRST EDITION

Designed by Bonni Leon-Berman

"Sympathy for the Devil," words and music by Mick Jagger and Keith Richards © 1968 (Renewed) ABKCO Music, Inc., 85 Fifth Avenue, New York, NY 10003. All rights reserved, used by permission of Alfred Music

Library of Congress Cataloging-in-Publication Data has been applied for.

ISBN 978-0-06-265620-9

19 20 21 22 23   LSC   10 9 8 7 6 5 4 3 2

For G, S, K, and P.
And for the donors, living and dead.
You are true heroes.

Contents

Note from the Author

The following book is neither a memoir nor a complete history of transplantation. I am not old enough to write a memoir, and a few excellent complete histories of transplantation exist already (and are listed in the bibliography). My goal is not to provide a chronological depiction of my coming-of-age as a surgeon, but rather, to use my experiences and those of my patients to give context for the story of the modern pioneers who made transplantation a reality.

The remarkable events that allowed mankind to successfully transplant organs between two individuals that are not genetically identical occurred relatively recently. In the early 1950s, the idea of transplantation remained in the realm of science fiction. By the late '60s, multiple organs were being transplanted, with a few poignant successes and many failures. True success with organ transplantation was realized in 1983, with the approval of cyclosporine. These accomplishments were achieved on the backs of a relatively small number of truly incredible people.

My own training began with four years of medical school at Cornell University Medical College in New York City. I then did my surgical internship and first year of residency at the University of Chicago Hospital and Clinics. After that, I spent the next three years performing transplantation research at Massachusetts General Hospital. I then returned to the University of Chicago for three more years of surgical residency. Thereafter, I came to Madison, Wisconsin, where I completed a two-year fellowship in organ transplantation. I have been in Madison ever since, performing organ transplants and running a basic science lab studying the immune system.

By illustrating what it took for me to practice transplantation, and by painting a picture, with the stories of my patients, of how the discipline has touched so many, I hope to highlight the incredible gift transplantation is to all involved, from the donors to the recipients to those of us lucky enough to be the stewards of the organs. I also will show the true courage of the pioneers in transplant, those who had the courage to fail but also the courage to succeed.

All the details in this book are historically accurate and factual to the best of my knowledge, with some minor patient details changed in a few cases to protect the identity of an individual, if requested.

Milestones in Transplantation

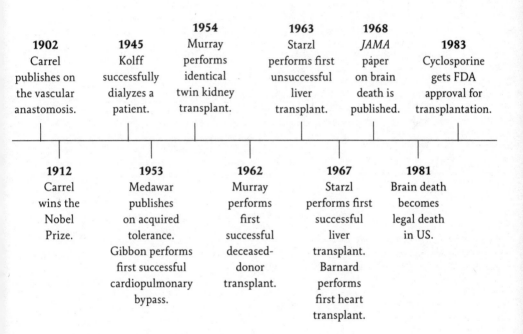

1902
Carrel publishes on the vascular anastomosis.

1912
Carrel wins the Nobel Prize.

1945
Kolff successfully dialyzes a patient.

1953
Medawar publishes on acquired tolerance. Gibbon performs first successful cardiopulmonary bypass.

1954
Murray performs identical twin kidney transplant.

1962
Murray performs first successful deceased-donor transplant.

1963
Starzl performs first unsuccessful liver transplant.

1967
Starzl performs first successful liver transplant. Barnard performs first heart transplant.

1968
JAMA paper on brain death is published.

1981
Brain death becomes legal death in US.

1983
Cyclosporine gets FDA approval for transplantation.

Part I

Out of Body

I have great respect for the past. If you don't know where you've come from, you don't know where you're going. I have respect for the past, but I'm a person of the moment. I'm here, and I do my best to be completely centered at the place I'm at, then I go forward to the next place.

—MAYA ANGELOU

We are not makers of history. We are made by history.

—MARTIN LUTHER KING JR.

| 1 |

A Perfect Organ

In a Small Plane over the Hills of La Crosse, Wisconsin,
September, 2:00 a.m.

While I'd been on planes many times, I'd never experienced the full power of a thunderstorm at ten thousand feet. The small King Air, a six-passenger dual prop, was bouncing around uncontrollably. Every few seconds, it would go into free fall and then hurl itself back up violently. The two pilots in the cockpit were hitting knobs and dials, trying to silence the various alarms that sounded as we rocked violently back and forth. It didn't help that our physician's assistant Mike, who had been on hundreds of flights in small planes before, was screaming uncontrollably, "We're gonna die! We're gonna die!"

Given that Mike was such a seasoned member of our team, I could only assume that this particular flight was going badly. When the pilots glanced back to see the source of the screaming and cursing, I could make out the fear in their eyes. I looked at the spinning altimeter and noted that our plane was popping up and down as much as a thousand feet at a time. Outside the window, the lightning was shooting horizontally. The rain was constant and loud, and I was sure I heard pieces of hail hitting the windshield.

IT WAS THE third month of my transplant fellowship at the University of Wisconsin. I hadn't chosen transplant surgery so I could fly through thunderstorms in the middle of the night over the fields of central Wisconsin. Hell, I'd grown up in New Jersey, spent most of my life in the Northeast, and had never known anything about the Midwest. I had been drawn to Madison because it is one of the best places to be a transplant fellow. I was learning how to perform kidney, liver, and pancreas transplants, and how to take care of these complicated patients while they waited for organs and then recovered from their surgeries.

One unique part of the discipline of transplantation is the procurement of organs from donors. While we do perform transplants, particularly kidneys, with organs from living donors, the majority comes from people who have just died. Rather than transporting donors, who typically remain on a ventilator, brain dead but with a beating heart, we send a team out first, to meet with their families to thank them for their gift and then to perform the surgery to remove their organs. We then take those organs back for transplant into waiting patients.

On this particular day, I'd received a phone call at around 5:00 p.m. telling me to come to the OPO (organ procurement organization) at 9:00 p.m., for wheels up at 9:30. The thirty-minute flight from Madison to La Crosse had been without incident. We arrived at the donor hospital at around 10:30. The donor was a young man (almost a boy) who had died in a motorcycle accident. That detail is easy to remember, as Wisconsin, being the land of Harley (not to mention a state where wearing a helmet is frowned upon), produces a never-ending supply of donors who've died in motorcycle accidents. In the winter, it's snowmobile accidents, the snowmobile being the vehicle of choice for bar hopping in the evenings—which sounds like fun but is also incredibly dangerous, given the power of those machines.

After we examined the donor at the hospital in La Crosse, confirming his identity and blood type, and went over the paperwork, including the declaration of brain death, we met with his family.

This continues to be one of the most difficult and, at the same time, most rewarding aspects of my job. No matter how tired I am, the interaction with the donor family always reminds me how wonderful and cathartic the donation process is. These people are going through the worst experience of their lives, as most donors die far too young and unexpectedly. Often, the family members have not even had the opportunity to say good-bye. Perhaps the one positive notion that family members can hold on to is this: with this ultimate gift, their loved one will save the lives of, and live on in, as many as seven other people. Their gift of life will be a legacy their families can cherish amid the brutal pangs of loss they have to endure.

We have a picture in our transplant unit of a mother whose teenage daughter died in a tragic car accident. This young girl saved at least seven lives. Some years later, the mother met the heart recipient at a transplant picnic we sponsored, and a picture was taken of her using a stethoscope to listen to her daughter's heart beating in the chest of the man she had saved.

This family tonight in La Crosse was no different. They asked how and when they could possibly make contact with the recipients, a process that we can help facilitate down the road if all parties agree. Then, once all their questions were answered, they said their final good-byes.

Once the donor was transferred to the operating table and prepped, we scrubbed in and placed the sterile drapes. At this point, all emotions from the encounter with the donor's family were pushed out of our minds. We had a job to do: to get all the transplantable organs out and flushed so that they would spring to life when placed in their new "owners." Our team, which had come for the abdominal organs, was not the only one in the operating room that night; there were two others: heart and lung teams, waiting to take their respective organs. We stood around the operating table, separated by the patient's diaphragm. They focused on the chest, and we focused on the belly.

I took a scalpel and made a long incision from "stem to stern," or from the notch at the bottom of the neck down to the pubis. As I dissected through the tissues and entered the belly, the cardiac team took a saw and began opening the chest. I quickly grabbed a malleable (a long, bendable steel retractor) and held it in front of the liver, to make sure they didn't get a little careless with the saw and injure this beautiful organ.

There is a natural conflict between a cardiothoracic team and an abdominal team. We all realize the importance of the incredible gifts the donor is giving, and we are all the stewards of these organs. At the same time, the procurement team always gets blamed for anything that goes wrong with the recipient operations that follow.

"Why is the upper cuff of the liver so short?"

"Why didn't I get more vena cava below my heart?"

We're all trying to bring back the best organs we can. So, everyone protects his turf.

I think about operations in steps. Step one: open the belly. Step two: mobilize the right colon and duodenum, and expose the aorta and vena cava. Step three: loop the aorta to prepare for cannulation (i.e., the insertion of a plastic tube into an artery that will allow us to flush the blood out).

That night, we got through our steps, which included freeing up attachments to the liver and separating the liver from the diaphragm and retroperitoneum. We dissected out the porta hepatis, identifying the hepatic artery and the bile duct. We divided the bile duct, letting the golden bile pour out into the abdomen. Then we cleared off the portal vein. Next, we mobilized the spleen and exposed the pancreas. As we neared the end, we identified the renal veins and arteries, which lead to the kidneys.

By now the cardiac team had scrubbed out and was standing behind us anxiously. Our portion of the operation is always much more involved than theirs, and as usual, they were constantly asking us when we would be ready. In their defense, their recipient surgeons (often hundreds of miles away) typically have already taken their patients to the operating room and begun opening their chests and getting them ready to be placed on bypass for removal of their sick hearts or lungs.

Finally, we were ready. We placed our cannula in the aorta. The cardiac team then placed a cross-clamp on the aorta and started infusing cardioplegia solution (which causes the heart to stop beating). Then they got ready to cut into the vena cava right before it entered the heart. (We made sure to protect as much vena cava as we could from those bastards. They didn't need it for their transplant, but we did for ours.) Once they cut it, the blood started to well up and out of the chest cavity. We started our flush through the aorta and then placed a second cannula in the portal vein. In poured the cold "University of Wisconsin solution," the wonderful solution, invented at our own institution, that preserves the organs and helps make all this possible.

The blood turned clear as it flushed out into our suction devices. We then poured buckets of ice into the abdominal cavity. Our hands began to cramp and ache from the ice as we held our cannulas in place. The good news was that, after a couple of minutes, the pain dissipated (as did all other feeling in our hands). The organs were cut out, flushed some more, and placed in bags.

Then we all went our separate ways.

That evening, I called Dr. D'Alessandro on the way out and told him we had a perfect liver. Of course, he was sleeping soundly in his bed. He would direct the OR team back in Madison to take the recipient patient to the operating room and start removing the old liver.

We took a cab back to the airport. It was about 1:45 a.m. at this point. We were all exhausted, but also filled with the satisfying feeling that always accompanies an operation gone well. The added bonus was that our cooler was filled with four organs that would go into three separate patients—a liver, a kidney, and a combined kidney and pancreas (called "simultaneous pancreas and kidney," or SPK). At the airport, we walked out onto the tarmac, where the pilots were waiting.

FOR SOME REASON I remember this vividly, even though it was more than ten years ago. It felt windy and cool that morning, quite different from the stifling summer weather we'd experienced when we landed a few hours before. There was the unmistakable feeling that a storm was coming.

The pilot turned to me and asked if I thought we should go. We both looked over at the cooler with the sticker reading "Organs for Transplant." I mentioned to him that he shouldn't worry about the organs; thanks to UW solution, the carefully designed preservation solution that would allow the organs to be metabolically inactive, they could wait awhile. I could always call Dr. D'Alessandro and tell him to delay the recipient.

Instead, I asked the junior pilot if he thought it was safe. I say "junior" because he looked all of about ten years old.

"It should be," he said. I detected a slight tremor in his voice.

Not that convincing, but I agreed to go.

We took off, and everything seemed pretty smooth. But about ten minutes in, things started to get crazy.

As the plane bucked and the alarms sounded, I really did think this was it. I thought about my family, particularly my little girl, born two weeks before we moved to Madison for my fellowship. I was bothered by the idea that someone at my funeral would say I'd died doing what I really loved. That's bullshit. There is really no great way to die, certainly not in a stupid little plane in the middle of the night.

We finally got through the storm, and as quickly as it started, it stopped. The pounding rain and the turbulence subsided, the plane settled, and we sat in silence for the last five minutes of the flight.

After we landed, I asked one of the pilots how commercial planes could possibly be landing in this weather. He said, "Oh no. The airport is totally closed, only open for emergencies." I remember feeling somewhat pissed about this, but in a way, what we had just done qualified as an emergency.

I OPENED THE bag with the liver and dropped it into the sterile bowl filled with ice. I was in the operating room back in Madison, and Dr. D'Alessandro and my co-fellow Eric were almost done with the hepatectomy, or removal of the diseased liver.

The donor liver was truly a perfect organ. I cleaned the extra tissue off it and meticulously tied off all the small vessels that came off the cava (though, of course I would still be blamed for any bleeding they got into after reperfusion). I then separated the pancreas (which we'd also use) from the liver, making sure not to injure either and to leave enough portal vein and artery for both transplants. I placed the pancreas in its own bag, which I would bring back down to the "lab." This organ would be prepared and transplanted in the morning into a type 1 diabetic, along with one of the donor kidneys we'd just procured. The other kidney would go into a different recipient. In two other states, two different patients were receiving the heart and the lungs from our donor in La Crosse.

I never cease to find this remarkable.

Once the liver was ready, I brought it into the recipient room, where the team was waiting. When they saw me, Dr. D'Alessandro took the Klintmalm clamp and placed it on the last remaining attachment to the liver, the hepatic veins going into the vena cava. He cut the recipient's liver out. I watched over his shoulder.

There is no more amazing sight in surgery than the abdomen once the liver has been removed. The vena cava—the large vein that brings blood from the legs back to the heart, which is normally enveloped by the liver—is fully exposed, coursing from bottom to top, and there is a huge, empty space around it. It is an unnatural but weirdly beautiful sight.

Dr. D'Alessandro took the new liver and started sewing it in—in steps. Upper cuff first. Then portal vein. Then flush. Then reperfuse. The liver pinked up and looked beautiful. Everyone looked happy.

Then Dr. D'Alessandro mentioned that I needed to go. There was another procurement, up in Green Bay.

2

Puzzle People

Madison, Wisconsin

My kids love to do art projects. They sit at the kitchen table and draw, cut, and glue princesses and animals and houses. The projects go on for weeks, cluttering up multiple rooms in our house, but in the end, the kids get a real sense of satisfaction as they play with their creations—until they are ready to move on to the next project.

My projects are my patients. Each one requires something cut out, glued in, or fixed up until it's time for me to move on to another. Cindy was a particularly memorable "project." When I first did a liver transplant on her, she was gravely ill, probably within a day or two of dying. I had gone to bed early the night before. At around two o'clock in the morning I got the phone call. I answered it on the first ring because, when I'm on call, as I was that night, I sleep with one eye open.

It was one of our coordinators for organ offers, Pamela. "We have a liver offer. It looks like a good one. He is a forty-four-year-old male, died of a drug overdose. Twenty minutes of CPR. Perfect liver numbers." Pamela spent the next five minutes giving various details about the donor's stability, previous medical history, and other lab values. I half-listened, partly because as long as the liver looked good, we were going to use it.

I asked Pamela who'd come up first for the liver.

"Cynthia R. MELD forty. Should I have the coordinator call you?" The MELD (or "Model for End-Stage Liver Disease") score predicts how sick a patient's liver is, and how likely she is to die without a transplant. A scoring system based entirely on lab values, the MELD score determines where a particular patient's name will fall on the transplant list. The scores range from 6 to 40. When your score is below 15, it typically means your risk of having a bad outcome during a liver transplant outweighs your risk of dying without a transplant, and we typically will not proceed. As the score gets higher, it means your liver is becoming more dysfunctional and you are at greater risk for dying without a transplant. Allocation of livers is based entirely on risk of waitlist death, with no consideration of quality of life, ability to work, or some prediction of your likelihood of returning home or to a "valuable" life afterward.

So begins the round of endless phone calls involved in coordinating every transplant, from identifying the potential recipients, who can be at multiple programs around the country; to bringing them into the hospital and making sure they are healthy enough to receive the organs; to running numerous tests on the donor to rule out infection risks; to running tissue typing to identify blood type and genetically match the donor and recipient; to setting up OR times at both the donor and recipient hospitals; to getting planes ready to fly all the donor surgeons and teams to the donor hospital; to, of course, making sure the donor family is comfortable with the timing so they can say good-bye to their loved one and talk to the transplant team. Each time there is a hiccup, and the timing has to change, all this has to be reset.

At 3:15 a.m., the phone rang again. (This second call is when I focus on the recipient.) Jaime, the transplant coordinator who focuses on our patients before and after transplant, gave me some more information on Cynthia, who goes by "Cindy." She had been admitted multiple times over the last few months. She had recently been treated for pneumonia and had spiked a fever the day before. She had gone into renal failure during this hospitalization and was now on dialysis. She was obtunded (i.e., confused from her liver failure to the point of almost being in a coma) and yellow as a banana. Her blood wasn't clotting at all, and she was oozing from her gastrointestinal tract (in her bowel movements), her nose, and around her IV lines. She was getting blood transfusions every day.

I trust Jaime, but given the severity of Cindy's illness, I decided to do my own chart biopsy. I turned on my computer and maneuvered through all the firewalls to log into the hospital system. I kept Jaime on the phone through this, since I might have to decide to call in a backup patient in case Cindy seemed too sick.

This crazy concept of calling a backup in case I deem Cindy too sick and skip her highlights the emotional challenge of being on the liver transplant list. When you are waiting for a liver, you want to be as healthy as possible going into this massive operation, but at the same time, you need to get sicker to get the liver, but not too sick that you get passed over when the time comes. When I evaluate an offer, I need to decide if I can get the patient through without killing her in the OR, fully aware that if I skip someone because she seems too sick, I am likely signing her death warrant.

Nowadays, we are willing to push things pretty far. I will take patients with breathing tubes, renal failure, fevers, and on medication to support their blood pressure. I will take patients who are having active GI bleeds, who have tumors growing in their liver, who have blood vessels that are clotted, and who may need bypasses of these blood vessels or some other heroic measure. But if I think someone is too sick, I'll have the backup recipient brought to the hospital, where the coordinator will tell this person that he will get the liver only if something happens to the person it is intended for. What must that be like for a patient—driving to the hospital in the middle of the night, getting prepped for surgery, even being wheeled down to the pre-op area, knowing he will receive this gift of life only if the intended recipient dies? It's horrific to contemplate, but from my point of view, I don't want to waste a healthy liver.

I looked over Cindy's data. I felt as if I knew her at this point, having gone through her history, labs, films. I have seen her digital insides, have examined her lungs and her liver and her spleen and her bowel and her blood vessels. If I saw her on the street, I wouldn't recognize her, but if I looked into her open abdomen, I would know her immediately, from her shrunken liver to her large spleen to her massive varices (big, swollen veins), which are carrying blood in the wrong direction (because of so much resistance to flow caused by that shrunken liver) and led to her GI bleeding, confusion, kidney failure, and now her imminent death.

I told Jaime it's a go. Let's not bring in a backup.

I finally met Cindy and her family at 4:30 p.m. I would come to know them quite well, particularly her daughter, Ally, and husband, Michael. I could see how much they loved Cindy, and how worried they were. I talked to them about the surgery and told them how sick she was. I told them the donor organ looked like a good liver. I went through some data—*x* percent chance of this, *y* percent chance of that, the possibility of bleeding, a bile leak, the clotting of blood vessels, organs being injured, the liver not working. But the questions they cared about most I couldn't answer.

First, I couldn't tell them much about the donor. We avoid giving too much information about donors, since it would be too easy to figure out their identities on the internet. And of course, Cindy's family wanted to know when the operation might take place. I had no idea. The various coordinators were busy trying to place all the organs. We had a good brain-dead donor (but with a heartbeat), so that meant being able to place the heart, lungs, liver, kidneys, pancreas, and maybe even the small bowel and skin, bones, and eyes. Some of the recipient surgeons involved probably wanted more tests done on this donor—a cardiac catheterization, an echocardiogram, a liver biopsy (which we'd requested), a bronchoscopy. That meant the donor would be wheeled down to the cath lab, where a cardiologist would stick a needle in his groin and snake a catheter in his heart to shoot pictures of his coronary arteries; another doctor would stick a needle into his liver for a biopsy; and a third would send a scope down into his lungs to look at his airways.

My phone rang just before midnight. Pamela again, back on her shift. An OR for the donor had been booked for 1:00 p.m. tomorrow. At 3:00 a.m. Pamela called again. The OR time had been moved to 3:00 p.m.

At 7:00 p.m. we were finally in the OR. The anesthesia team put Cindy to sleep. I sat in the room watching them put gigantic IV lines into her neck, to pour blood in when I began exsanguinating her. I noticed that her systolic blood pressure was starting at 60 mmHg, dangerously low considering I hadn't even started the bloodletting. I watched them dial up the blood pressure meds. I considered whether I should call in a backup recipient now, as the likelihood of Cindy not making it had risen a little bit. *Nah, forget it.* Our team was packing up at the outside hospital, a thirty-minute flight away. Our fellow had sent a picture of the donor liver to my phone; it looked perfect. *I'm putting this thing into Cindy. It's hers now. If she gets buried, it will be buried with her.*

At 8:15 p.m. we finally made the incision. My second-year fellow Emily sliced through Cindy's skin. Everything was bleeding. This was not surprising, given that she had absolutely no clotting factors in her body and was already bleeding from every IV site and orifice. But she was not Cindy to me anymore. I no longer thought of her life, her family, whether she was male or female, young or old. I don't think I would have been able to do to her what I was about to do if I thought about that. I had seen her films, had a mental image of what everything in there should look like, but now I needed to put together the puzzle.

A liver transplant can involve anywhere from a thousand to a million pieces. Despite the exquisite quality of CT scans and MRIs these days, you never quite know what a liver transplant, or any operation, is going to be like until you start. But shortly after opening, you have a pretty good idea. If the diseased liver is shrunken and mobile, and you can reach in and pull it up right away, you know it won't be that difficult to get it out. If it is stuck to the tissue around it from years of inflammation and damage, however, you know you're in for a battle. If you lose 2 liters of blood just cutting through the skin, you know you're screwed.

We got into Cindy's belly and sucked out 8 liters of beer-colored fluid—we call this fluid ascites; it bathes the organs in most patients with advanced liver failure. (I congratulated Steph, our scrub nurse, for guessing 7.5 liters—the closest to the actual number without going over. Good, clean OR humor.) We put our retractors in and looked at the liver. I could tell right away this was going to be bloody but not that bad. We got two suctions ready to suction the ascites and blood we were going to be swimming in throughout the operation. Emily and I had both put knee-high waterproof "booties" over our OR clogs, so we wouldn't be standing in soggy socks by the end of the operation (something we have all learned the hard way).

The surgery itself went quite well. Cindy's beautiful new liver worked right away, and we were able to stop the bleeding without too much trouble. We finished at around 3:30 a.m.—a little long for some surgeons, but I believe in taking my time and making sure everything is perfect before I leave. I went downstairs to talk to the family, leaving Emily and "anesthesia," as we refer to the anesthesia team in the OR, to move the patient to the ICU and to complete all the never-ending paperwork. I told the family the case had gone well. I mentioned that Cindy's pressure was pretty saggy throughout, but I was hopeful that would correct itself over the next day or so. She would go to the ICU with a breathing tube in. She was critically ill, but I thought she would be okay. They asked me if her kidneys would recover, and I said I hoped so.

EMILY CALLED ME at around 8:00 a.m. on post-op day five. "Hey, Josh, it looks like there is bile in Cindy's drain." Damn. My stomach immediately jumped into my mouth. Every time something goes wrong with one of my patients, I get this incredibly awful feeling, something like guilt mixed with nervousness mixed with depression. Bile can't be good. It had to be leaking from where we'd sewn her bile duct together. The two ends of the duct had seemed fine in the OR, but she had been so unstable over the last couple of days that perhaps her low blood pressure had caused the ducts to fall apart. Bad blood flow can lead to bad healing.

In the shower, I pictured the operation we would need to do—most likely a Roux-en-Y hepaticojejunostomy. In other words, we would divide her small bowel, pull one end up to the bile duct, sew that onto the bowel, and then plug the other end back into the bowel so it looked like a Y.

Now the guilt was seeping in. You really have one shot at getting surgery right on a sick patient, particularly one who is on immunosuppression (i.e., drugs to prevent the immune system from attacking the new organ). Once you have a complication, you're backpedaling.

I could already see that Cindy would now be hospitalized for months, would likely get numerous infections, have prolonged intubation, have an open wound, need various antibiotics, probably grow fungus out of her belly, get line infections and deep vein thromboses (DVTs), and probably never come out of renal failure. Nice thoughts.

I drove to the hospital and headed directly to the ICU. Cindy's drain looked like shit. I don't mean it looked bad. I mean it actually looked, and smelled, like shit. I wasn't sure what I'd screwed up, but I couldn't help wondering if another surgeon could have avoided this. Emily got the OR ready, and we rushed Cindy off for surgery. I was super anxious until we got her into the operating room. As a surgeon, when you have a complication, you're dying to fix it. Waiting to go to the OR is agonizing, and sometimes it seems like everyone is putting up barriers to your doing so—missing paperwork, delayed lab results, absent staff.

We opened Cindy up and scooped out what seemed like a liter of poop. We saw that the liver looked great (other than being poop-stained) and the blood vessels were fine, as was the bile duct. As we looked around, we found a large hole in her right colon. I had no idea how that got there. Maybe it was from a retractor; maybe it was from her low blood pressure and high-dose steroids. It wasn't directly my fault—but did that make any difference?

Emily and I removed Cindy's right colon and gave her an end ileostomy and long mucous fistula that she would keep for the next year. In other words, we pulled the end of her ileum (small bowel) right through her abdominal wall so that her stool would come directly out into a bag; we also pulled the disconnected colon out as well as a double-barreled ostomy, so it couldn't leak into her belly. After that repair, Cindy had a pretty tough course—a three-month stay in the hospital; an open, gaping wound; several readmissions; rounds of nursing home care. But she finally got better and made it home. And we were able to reconnect her bowels so she could poop like the rest of us again.

Her family was with her every step of the way, and it was definitely tough on them, but they weren't done giving. Cindy's kidneys never recovered. She was going to dialysis three days a week, four hours at a time. It is a miserable existence, but it was keeping her alive. The good news was this was something we could fix. We just needed a kidney. And immediately, her daughter, Ally, stepped forward to give it. Once that kidney was in there, she would be as good as new, ready to get on with her life.

This is why I love the field of transplant. Since I began taking care of sick people, I have noticed that one of the hardest things about getting sick, really sick, is that you are separated from the people you love. Even when families are dedicated to the patient, illness separates the well from the sick. The sick suffer alone, they undergo procedures and surgeries alone, and in the end, they die alone. Transplant is different. Transplant is all about having someone else join you in your illness. It may be in the form of an organ from a recently deceased donor, a selfless gift given by someone who has never met you, or a kidney or liver from a relative, friend, or acquaintance. In every case, someone is saying, in effect, "Let me join you in your recovery, your suffering, your fear of the unknown, your desire to become healthy, to get your life back. Let me bear some of your risk with you."

I saw Cindy in my office on a Tuesday in October, about a year and a half after her liver transplant, the day before I was planning to perform her kidney transplant. She was with Ally. Cindy tearfully asked me what the chances were that the operation would be successful.

"Of course it's going to work," I told her. I didn't say this out of a surgeon's narcissism. The reality is that she was getting a young, living-donor kidney from a healthy donor, all the immunologic tests we had performed indicated no evidence for risk of early rejection, and the procedure has become quite commonplace. I just needed to put the last piece of the puzzle in place, and she would be on her way.

It really was that simple.

How did we get here? How is it possible that we can take organs from someone who has just died, plug them into someone who is in the process of dying, and have those organs suddenly start working? Livers begin making bile, kidneys start peeing on the table, pancreata start secreting insulin and regulating blood sugar, hearts start beating, lungs start breathing. It has all become so straightforward and predictable, but it wasn't always like this. There was a time when sane people thought transplant was a pipe dream, something that could never happen.

Lyon, France, June 24, 1894

Marie-François-Sadi Carnot, the popular French president, had just given a speech at a banquet in Lyon, and was back in his carriage, when a man ran at him from the crowd. Sante Geronimo Caserio was a twenty-one-year-old Italian anarchist who had made up his mind to kill the president. He'd purchased a knife. He'd studied the program for the president's visit to the city. When the perfect moment arrived, he jumped onto the president's carriage and stabbed him. Carnot was taken to the town hall, where prominent local surgeons examined him. They probed his wound, and he briefly came out of his unconscious state, exclaiming, "How you are hurting me!" Shortly thereafter, he died. An injury to the portal vein was identified as the cause.

One can only imagine the chaos and emotions that this assassination inflicted on the people of France, heightened by the complete inability of surgeons at that time to offer Carnot anything resembling treatment. His murder had a major impact on one young student, Alexis Carrel. An extern in Lyon (the equivalent of a medical student), Carrel wondered whether he could somehow improve the management of these types of injuries and decided to become a surgeon. He was a natural surgeon, ambitious, driven, and hungry for fame. He reportedly told people that doctors should have been able to save Carnot, that there should be a way to sew severed vessels back together. Surgeons of the day thought the idea was crazy.

In 1901, once he finished his initial training in surgery, Carrel obtained space in a lab with access to surgical equipment and dogs. His focus was on designing a method to join together two blood vessels. It is hard to imagine that surgery ever existed without this, but at the time, there was no inkling of peripheral vascular disease, no real understanding of atherosclerotic plaques, no consideration given to operations on the heart, and most people didn't live long enough to develop these types of problems anyway. While vascular injuries were seen secondary to battle wounds or trauma, the standard management of these injuries was to try to ligate (tie off) whatever might be bleeding; this remained the practice well into World War II. The main vascular issue surgeons saw in those days involved aneurysms (or the outpouchings of arteries), which nowadays are associated with smoking and atherosclerosis. Back then, though, these were often secondary to syphilis. If an aneurysm was found before it ruptured, causing certain death, surgeons would ligate the artery. Mortality was high, but not that much higher than the outcomes seen with virtually any abdominal operation in those days. Thoracic operations weren't even attempted.

Carrel recognized three things: First, he needed to find better needles and thread to sew vessels together, thus minimizing injury to the inner wall (intima); the needles in use then were causing clots to form at the needle holes. Second, he wanted a technique that would protect the intima more than he could do just by improving his suture material. Third, he needed to find a setup that would allow the repair to be done quickly, as he knew that clamping the vessel for too long would inevitably lead to clotting. Knowing that the needles and thread available for surgery were woefully inadequate, he visited a local haberdashery in Lyon to obtain finer material, which included straight needles and fine cotton thread.

In addition, legend has it he took embroidery lessons at the home of Madame Leroudier, a world-famous lace embroideress, and practiced sewing with these needles on paper until his technique was perfect. He placed paraffin jelly over the needle and thread, to allow it to be pulled through the tissue more easily, and in 1902 published a paper describing his findings.

Alexis Carrel has always been described as a gifted surgeon. Most surgeons fall somewhere on a bell curve of inherent surgical skill, which is adequate to obtain good outcomes, even in technically complex cases. That said, there are natural surgeons whose hands are so good that within just a few minutes of working with them, you can tell they are off the bell curve. There are no wasted motions, the moves are so efficient, every stitch is perfect, and their instincts are unnaturally good. Carrel was in this group; he was a physical genius.

In addition to his physical skill and adoption of better equipment, Carrel was passionately committed to organ transplantation, a discipline that depends on sewing together blood vessels to supply the new organ. This is rather remarkable, given that organ transplantation was still in the realm of science fiction at this point, with a few sporadic attempts that were universally followed by rapid failure.

Carrel presented his results at local scientific meetings in Lyon, with generally good reception. He hoped that his description and demonstration of the vascular anastomosis (or reconnection of blood vessels), along with some follow-up experiments in which he sewed the carotid artery (the main artery in the neck that goes to the brain) end to end to the jugular vein (the main vein in the neck that drains the head) in dogs, would help him secure a junior faculty position. The artery-to-vein experiment was also well received, and the technique was presented as a possible treatment for strokes or general mental decline by increasing oxygenated blood flow to the brain. We know now that this would have no beneficial effect, but it was a concept Carrel would explore over the next decade as a treatment for various failing organs. But as would become a recurring theme in the life of Alexis Carrel, some of his greatest achievements would be diminished by controversies he entangled himself in due to his diverse interests and beliefs. One local paper quoted him voicing his belief in supernatural healing forces at the shrine at Lourdes. Carrel had a mystical belief in the supernatural, and felt there existed powers that could allow the healing of various maladies and diseases in a rapid fashion. This notion was met with ridicule, and he was passed over for a staff appointment. Feeling betrayed and stifled in Lyon, he decided to emigrate to North America. After a brief stop in Montreal, he was recruited to work with Professor Carl Beck in Chicago, both working at Cook County Hospital on humans and doing experimental surgery on dogs. It soon became clear to him that he had no interest in performing human surgeries. He also had a fairly low opinion of American surgeons. He described "the crowd of imbeciles and villains who corrupt the world of medicine . . ." and declared that "to be a medical doctor in the United States is the lowest form of business." An opportunity arose at the University of Chicago, where he would not have to take care of human patients at all, and facilities were available for animal surgery. There he met Charles Claude Guthrie, a physiologist and researcher whose lab was performing dog surgery. The two men worked together for two short stints of three to four months, yet in that time, they published ten papers in American journals and almost twice that many in international ones. This productivity was certainly driven by Carrel's appetite for fame and recognition as well as a sense of the competition that was arising in the field of vascular reconstruction and even organ transplantation in animals. It is truly remarkable how many different operations use the vascular anastomosis Carrel and Guthrie considered and described in that short period. These included connecting the femoral vein to the femoral artery in the leg of a dog (to improve blood flow to the leg); updating Carrel's original technique of the vascular anastomosis, taking full-thickness bites through the entire wall of arteries rather than just the outer layer; performing vascularized thyroid grafts, by either removing the organ from and replacing it in the same animal or transplanting an organ between different animals; and attempting multiple kidney transplants. Encouraged by their success, they also transplanted a canine heart from the chest of one dog to the neck of another (which beat for as long as two hours) and made attempts at transplanting both hearts and lungs (which invariably failed). In 1906, they published a paper on the use of the "Carrel patch," which involves cutting a vessel out along with a rim of aorta to make it bigger, a technique we still use today to perform organ transplantation.

This may have been the most critical year in Carrel's illustrious career, for two reasons. First, he focused singularly on mastering the technical demands of the vascular anastomosis. This attention, obsession even, to getting the technique perfect with repetition and focus was crucial. As a transplant fellow at Wisconsin, it took me two years of sewing in organ after organ, day and night, before my muscle memory had developed to the point where I didn't have to think at all when I sewed. When you first start sewing together vessels, you must constantly keep in mind whether you are inside or outside each vessel wall, and you are never sure how big a bite to take with your needle or how far to advance. At some point in your training, you load the needle on your needle driver and turn your body without even thinking about it, and something that originally might have taken thirty minutes to an hour becomes a ten-minute exercise.

Of course, when I operate now, I have a scrub tech assisting me; a resident or fellow across from me; sturdy and complex retractor systems that hold everything out of my way; powerful overhead lights and a headlight to illuminate the field; super-sharp, well-designed, fine needles with even finer coated sutures that glide through the tissue; and spring-loaded needle drivers that I can operate with just my fingertips. Carrel had none of this.

The second reason 1906 was such an important year for Carrel has to do with his obsession with publishing. Some of the publications from this year remain relevant to the practice of medicine today, and his predictions about the application of the procedures he discussed in them, particularly in the field of transplantation, are shockingly prescient. By far my favorite work of his has to be "Successful Transplantation of Both Kidneys from a Dog into a Bitch with Removal of Both Normal Kidneys from the Latter," published in the premier journal *Science*, no less.

The other avenue Carrel started traveling down that year was interacting with the lay press. Although the practice was rare, and looked down upon by scientists and surgeons of the time, Carrel developed relationships with members of the press and leaked sensational information about his experiments to them. He also shared his techniques with renowned surgeons of the era. When a surgical society was convened in Chicago, Carrel had the opportunity to demonstrate his vascular anastomosis in a dog to more than twenty prominent surgeons, including the rising star Harvey Cushing. Cushing was at the Johns Hopkins Hospital at the time, working with the great William Halsted, perhaps the father of American surgery. On April 23, 1906, Carrel traveled to Baltimore to present his findings at the Johns Hopkins Hospital Medical Society. In the audience could be found some of the premier surgeons and physicians of the era, including Halsted, William Welch (one of the founders of Johns Hopkins Hospital), and William Osler (a Hopkins founder and often considered the father of modern medicine).

Carrel spoke that day of his vascular anastomosis, the use of vein grafts to replace sections of arteries, and the importance of asepsis (the absence of bacteria or viruses) in outcomes of vascular anastomoses. (Joseph Lister had been pushing the importance of this in surgery starting in the mid- to late 1800s, but it certainly was not accepted practice, and hand washing and the use of gloves were not yet the standard of care.) Finally, he spoke of his experience with organ transplantation, its possible future applications, and how these surgeries seemed to fail after a week for unknown reasons. While he certainly did not call this "rejection," or appear to have much understanding of the immune response at this time, he did refer to possible inherited factors and discussed his plan to "perform a series of similar operations on pure bred animals" to understand this failure better. He also stated that "we intend to try and immunize the organs of an animal against the serum and organ extracts of another . . . The transplanted organ must be prepared to support the serum of the animal on which it is to be grafted." As a practicing transplant surgeon with a lab focusing on the immune system, I am pretty blown away by the topics Carrel spoke about and the predictions he made. The fact that he did most of this work in such a short time is also mind-boggling.

Carrel got a similarly positive response from the Hopkins crowd, and they tried their best to secure his appointment to their institution. But the infrastructure to support medical research was just starting to be put in place in America, and the lab space at Hopkins was just being built. Also, at the same time, another offer presented itself. Efforts were then under way to start institutes for medical research modeled after some of the great ones in Europe, institutes that could put American medicine on the map. The National Institutes of Health did exist at this point, but functioned essentially as a small laboratory and didn't start giving extramural grants for research until after World War II. Instead, two fabulously wealthy businessmen, John D. Rockefeller and Andrew Carnegie, decided to spend large portions of their fortunes to support medical research. In September 1906, the first director of the Rockefeller Institute, pathologist Simon Flexner, was successful in attracting Carrel to the gleaming labs in the newly built institute on the banks of the East River in New York City.

At Rockefeller, Carrel's most remarkable experiments were surgical in nature, involving blood vessel surgery and transplantation. In the field of transplant, he did virtually everything. He performed vascularized transplants of spleens, thyroids, intestines, and an ear (supplied with blood by the external carotid artery). He performed numerous leg transplants between dogs, sewing the blood vessels together and nailing the bones in place. Perhaps his most important transplants were kidneys. He first perfected the autotransplantation of kidneys in dogs (taking the kidney out and then transplanting it back into the same dog). He then moved on to transplants between two different animals. He thought about his occasional longer-term successes and came to the conclusion that something about the close relationship between siblings could make grafts last longer. And most impressively, Carrel grasped what might have been the next step in making transplantation a clinical reality: he considered the idea of manipulating the donor graft prior to transplantation or applying some sort of conditioning to the recipient. It was this work that led to his Nobel Prize in 1912, "in recognition of his work on vascular suture and the transplantation of blood vessels and organs."

Shortly after this, James Murphy, working in the Rockefeller lab of future Nobel laureate Francis Peyton Rous, published a paper showing that lymphocytes (cells of the immune system) could "reject" tumors and stop their growth when they were transferred to other chicken embryos. This was essentially the first explanation of transplant rejection, and Carrel recognized this. Furthermore, Murphy showed in mice and rats that by either irradiating them or treating them with the chemical benzol, lymphoid tissue would be damaged, lymphocytes (and hence immune function) would be decreased, and tumors could be transplanted and survive in these animals. Carrel took the next step, in his mind, and considered that either radiation or chemicals such as benzol could be given to transplant recipients to extend the lives of grafts.

In 1914, when he went off to France for his summer vacation, World War I broke out. Tragically, this concept of recipient treatment and the role of lymphocytes in graft failure would essentially be lost until the 1950s.

When Germany declared war on France, Carrell was thrilled, as he loved the military and felt France had gone soft. War was just what was needed to cleanse the soul of the French people. Carrel had a particular interest in the management of wounds and wound healing, and he formed a relationship with an American chemist by the name of Henry Dakin, whom Carrel tasked with coming up with a strong antiseptic solution that could be used to wash out battlefield wounds. With it, Carrel devised an intricate and painful wound-perfusion system, which eventually fell out of favor due to its complexity. Dakin's solution, with some modification, is still used in the management of open wounds today.

After the war, Carrel spent an additional twenty years at Rockefeller. In the lab, he moved away from surgical work and turned to cell culture and made some minor advancements in that field, which he portrayed to the press as major breakthroughs. Ultimately, some of these breakthroughs would be exposed as fraudulent. In addition, he wasted time and money on expensive experiments that were badly designed. For example, he conducted a large mouse experiment looking at the role of diet and environment in the development of cancer, but it was poorly controlled and without any testable hypotheses. But overwhelming his life at this time was a growing interest in eugenics and his relationship with Charles Lindbergh, the famous aviator and eugenicist.

From the turn of the twentieth century up until World War II, eugenics was widely popular in the United States and Europe. More than three hundred major universities offered the study of eugenics as part of their curriculum, and the scientific community considered it a legitimate science. A short list of famous people who subscribed to it includes Theodore Roosevelt, Alexander Graham Bell, John D. Rockefeller Jr., H. G. Wells, Winston Churchill, John Maynard Keynes, Woodrow Wilson, Henry Ford, and Francis Crick.

So, what does this all have to do with Carrel? In the 1920s and '30s, Carrel embraced the concept that Western civilization was in a decline. He became so focused on his concerns about mankind that he dedicated himself to the writing of his opus *Man, the Unknown*, which supports a positive version of eugenics. He wrote extensively about the loss of "natural selection" in mankind and a need to develop the strong.

The book was released in 1935 and became a best seller. Many of its themes, particularly the general decline of Western societies, the potential for improving living things with selective breeding, and getting rid of criminals and those deemed insane, were popular and mainstream. Carrel was at the absolute peak of his popularity in these years, until his retirement and return to France in 1939, where he was supported by the Vichy government.

If Alexis Carrel had followed his instincts, he very likely could have become the true father of transplantation and perhaps one of the greatest innovators in the history of medicine. If he had simply died or faded away shortly after the Nobel Prize he won in 1912, at the very least he may have been revered as one of the premier experimental surgeons of the twentieth century. Instead, as his research production dwindled and his relationship with Charles Lindbergh blossomed, Carrel focused more and more on the degeneration of mankind and how he could play a role in studying this in a scientific way. He always had a fascination with strongmen such as Mussolini and Hitler, and thought that in the 1930s, German society was taking a good approach to cleaning up its population.

Shortly after the liberation of France from the Nazis in August 1944, various rumors emerged that Carrel was under house arrest, was going to be tried as a German collaborator, or was on the run. None of this was true. He had fallen ill from heart disease, having suffered his first heart attack in 1943, and died in November 1944. Although he was not formally charged with any crimes, his name became strongly associated with Nazism, fascism, and anti-Semitism. His reputation was destroyed, and to some degree, many of his discoveries were lost. His transplant work was completely forgotten, with little mention made of it to this day.

Sir Roy Calne, the man who perhaps more than anyone moved

the field in the direction of chemical immunosuppression, had this to say about Carrel: "Alexis Carrel was a brilliant researcher, but not a very nice man." Despite that, his contributions with regard to the technical aspects of sewing vessels together and transplanting organs from one animal to another with initial graft function represent the first piece of the puzzle of organ transplantation.

Part II

The Making of a Transplant Surgeon

Only one who devotes himself to a cause with his whole strength and soul can be a true master. For this reason mastery demands all of a person.

—ALBERT EINSTEIN

| 3 |

The Simple Beauty
of the Kidney

In order to attain the impossible, one must attempt the absurd.

—MIGUEL DE CERVANTES

What does a kidney transplant look like, from the beginning? Assume the patient is asleep on his back on the operating table, prepped and draped. I first identify my landmarks on the abdomen, making a mark on the patient's belly with an erasable marker. I mark the rib cage, anterior-superior iliac spine (hip bone), and pelvis. I then put a mark two finger widths medial to the hip bone and draw a line that goes from the pelvis through this mark and up to the rib cage. The line will look like a curved incision that goes from the pelvis, up the abdomen, and ends just above the belly button. I usually do it on the right side, because the blood vessels are a bit more toward the front on this side. We can use the left as well, which is what we do for retransplants (when a first transplant fails). I now take the knife and cut through the skin along the line. I then use the Bovie electrocautery, which uses electric current to heat, divide, and cauterize tissue, to cut through the fat and down to the muscle layer. I then continue through the external oblique muscle, internal oblique muscle, and transversalis muscle and fascia. I actually divide these muscles, and even though the patient is asleep, the muscles contract in response to the electric current, jumping back and forth.

I now identify the peritoneum, that is, the "bag" that surrounds the intra-abdominal organs. The small and large bowel, stomach, spleen, liver, and part of the pancreas all sit inside this bag, and you have to cut through it to get to them. Other organs (kidneys, part of the pancreas, part of the duodenum, and big blood vessels including the aorta and IVC, or inferior vena cava) sit behind this bag, so you have to push it out of the way to get to them. In a kidney transplant, we don't typically remove the sick kidneys but, rather, place the new kidney lower down on the blood vessels that ultimately go down the leg. In the early days of transplant, the native kidneys often were removed, but eventually it became clear that this added more surgery with no benefit.

Trying not to make a hole in the peritoneum, I pull it out of the way, and behind it I start to see the iliac artery and vein, and the patient's ureter. I dissect out the iliac artery and vein (which continue down the leg). This involves using scissors, forceps, and the Bovie to divide fat, lymph, and small blood vessels off these structures. I tie off any small vessels that I think might bleed (which means I take a silk suture and tie these little guys just like you might tie your shoelace). This part of the operation can be really easy, but in someone who has had previous surgery, or who has a lot of atherosclerosis (plaque in their arteries) or is obese, it can be much harder.

At this point, I focus on the donor kidney. In most cases, I've gotten it ready while anesthesia is putting the patient to sleep. (Back-table surgery, as this is called, is very specific to the field of transplant. It takes a while to learn, because the organs can be held upside down and the blood vessels are empty and flaccid, making it easy for a surgeon to inadvertently cut them in half. In addition, every kidney is different. Some have multiple arteries, some have multiple veins, and some even have two ureters!) I meticulously clean the excess fat off the donor kidney and then have an assistant hold the renal artery toward me while I dissect up it, tying off all the little side branches. I do the same to the vein. In doing this, I take care not to injure the ureter and its blood supply. If there are multiple arteries to the kidney, I need to decide if I can sew them into the patient with one big Carrel patch that includes all the orifices. If two arteries are too far apart, I have to decide if I will "pants" them together—that is, make a slit down the lumen (the opening of the artery) and then sew them together, making one lumen out of two—or, alternatively, if I will just implant the arteries separately onto the iliac artery. I then flush the vessels with cold saline, stitching up any small holes I find when the saline squirts out of them.

It's better to do all this on the back table than in the patient. Surgery is all about predicting where something bad might happen and then doing something to prevent that from happening. I take my time on the back table, the most critical point of the case.

Now I am ready for implantation. I have the anesthesiologist give the mannitol and Lasix to protect the patient from the evil humors that are released when blood flow is restored to the kidney. I place a side-biting clamp on the iliac vein of the recipient, then cut a slit in the vein, making sure not to get too close to the clamp. Using angled scissors, I extend this slit to match the size of the renal vein from the donor kidney. I then bring the kidney up on the table. I get a 6–0 Prolene suture—Prolene is nonabsorbable, meaning it will be there forever—and put one stitch inside out through one end of the iliac vein and the other end inside out in the donor renal vein. I put a second one similarly in the other end, inside out in the iliac vein and donor renal vein. I like to put a third stay suture in the middle for retraction, again inside out and inside out. Then I drop the kidney down into the field and carefully tie the sutures, three to four knots in each. I have an assistant retract the kidney away from me, and I load one of the needles. I then run the suture from one end to the other, outside in, inside out, being sure not to catch the back wall. Once one side is done, I tie the suture to the one I had placed at the end—seven knots. I flip the kidney back to me and take out the stay suture. I then sew the other side, from one end to the other.

Now I turn my attention to the artery. I place clamps above and below where I plan to sew, taking particular care not to injure the artery by clamping too hard. I then make a small hole in the artery and use a circular punch to punch out a bigger hole—somewhere between four and six millimeters in diameter. (Although it can be smaller, I try to match it to the size of the donor vessel.) I get another 6–0 Prolene suture and, again, go inside out on the iliac artery and then the renal artery. I pull the artery down toward the iliac and tie it down—four knots. Then I pass the Prolene behind the artery and sew the back wall first, from the outside. Outside in, inside out—every stitch needs to be perfect. The first few stitches on this anastomosis can be a bit challenging and awkward, but once you have done it a few hundred times, it gets easy. No back walls, no dissections—it is critical to get all the layers of the artery when you sew this, so you don't raise a flap. (Arteries have three layers, whereas veins have one. If the inner, intimal layer separates from the outer two layers, it can become a flap, and over time blood can dissect into this layer and cause the artery to clot.) Then, the moment of truth.

I release the clamps and watch the newly transplanted donor kidney turn pink. A minute later, it squirts urine into our field and on our hands while we continue working. What a beautiful sight! I spend a few minutes drying things up, and then move on. I pull the bladder, which is really just a muscular bag—yet hard to find in a big, deep person—up into the field. I have gotten in the habit of having the nurses fill the bladder from below (through a catheter that goes through the urethra and into the bladder) with blue fluid. This way, I can stick a needle in any structure that looks like the bladder and be sure the syringe is sucking up blue fluid. When I successfully find the bladder and see my beautiful blue dye, I open it up and sew the ureter onto it over a stent. I then dry things up again. Then I close the muscle and fascia in two layers, making sure not to nail the peritoneum or its contents.

Again, everything needs to be perfect. It doesn't matter how tired or distracted you are, how many things might be going on with other patients or with your boss or your lab or in your personal life. It needs to be perfect. Otherwise, the patient will pay a huge price, the donor won't have given the gift of life, and you will be woken in the middle of the night by a shrill pager letting you know you've screwed up, it is your fault, and now you have to deal with it.

That's a kidney transplant. No big deal, but one of the best things we do in health care.

The kidney is an exquisite organ. I like to tell my residents that "the dumbest kidney is smarter than the smartest doctor." In a healthy person with a working organ, blood flows into the kidney and goes through an ingenious system of glomeruli—that is, circular tufts of thin blood vessels surrounding the tubules of the kidney. Across the kidney's membranes and structures, toxins, wastes, and electrolytes are filtered out into the tubules to be secreted as urine. Kidneys are also involved in controlling blood pressure and stimulating the production of red blood cells. It's amazing how a working kidney seems to know exactly what to do with fluids and reabsorption, whereas we doctors have so much trouble regulating fluid in patients, no matter how many labs and vitals we check.

In part due to that complexity, and the significance of this organ in a functioning body, up until the 1960s, patients with kidney failure would simply die. Chronic or intermittent dialysis, which is essentially renal replacement therapy (in which people are hooked up to a machine that filters blood), didn't become a reality until then. The futility of treating patients with kidney disease is what inspired Alexis Carrel to take the first step toward making transplant a reality, simply sewing a kidney from one animal into another. But many more steps would need to follow, all inspired by the desperation of watching young men and women die because they had lost the ability to urinate.

And so, it was fitting that it was a kidney transplant that started me on the path to becoming a transplant surgeon.

Cornell University Medical College, New York City,
Third Year of Medical School

When I started my third year of medical school, I still had the idea that I would be a pediatric oncologist—until I did a rotation on a pediatric service. It didn't take me very long to realize that that was not the field for me. The pace was slow, the parents were challenging, and I spent too much time doing well-baby checkups. I then did obstetrics and gynecology, and while I did enjoy this more, especially the excitement of delivering my first baby, it didn't take me long to realize I was not a budding gynecologist. My internal medicine rotation came next. I got to talk to patients, listen to their stories, and figure out what might be wrong with them, but I didn't like the pace or the chronicity of the diseases we would deal with. Everyone had diabetes, hypertension, or cancer, and we weren't going to cure any of these conditions. We would see twenty-five patients in the clinic a day, each one for only fifteen minutes, which wasn't enough time to address any of the multitude of problems each had. We would focus on one or two issues, adjust a few meds, prescribe some new ones, and move on, unsure that any of the lifestyle and medication changes we'd suggested would actually happen.

Next came surgery. My rotation started early in January, shortly after the holidays. I had come in at about 4:30 in the morning to do rounds—visit some patients, gather their labs, and write their progress notes (daily notes written in the chart of every patient). Other than the sound of the nurses and the few medical students on the surgical services gathering patient information, it was quiet in the hospital. The interns showed up at around 5:30, and the medical students updated them on the patients. At 7:00 a.m. we met in the cafeteria to present the cases to the chief resident. The medical students sat silently at the table, not daring to talk or eat. The third-year resident presented the patients to the chief resident, and, turning her sharp gaze on the interns, commanded them to clarify any details she didn't know. The chief listened, casually eating a donut.

I went to the operating room at 7:30 and scrubbed in on two elective cases, a resection (removal) of a rectal tumor and a liver resection, during which I was soundly criticized for my inability to cut sutures well. Then I did rounds again with the team, gathered more patient data, and went to a conference. At some point during the day, I was told I would be on call that night. This was good training for the next twelve years of my life. Don't make plans. You are expected to be available days, nights, weekends, holidays, and family birthdays.

At about ten o'clock in the evening, we took a patient to the operating room for a bowel obstruction. I had been both too nervous and too busy to eat that entire day. I started to feel dizzy. I remember looking at the patient's distended bowel, with its liquid contents sloshing around. I started to feel faint and was sweating profusely. Somehow, I gritted my teeth and got through it. As the case was ending, my resident told me to go scrub in next door—it was about 2:00 a.m. now—where they were starting a kidney transplant.

All I really wanted to do was go to bed, but I went. And it was amazing. Dr. Stubenbord was the transplant surgeon. I'll never forget the simple beauty of the kidney transplant, the feeling of wonder when the kidney turned pink. There was this remarkable sense in that room, in the middle of the night, with classical music playing and urine pouring onto our hands, that we were doing something miraculous. Someone who had just died had saved the life of someone he had never met, and we were the ones who'd helped make that happen. (Well, not me, really; I just watched.) How crazy was that? I wondered what other organs you could do this with. I really had no idea at the time, but I knew I wanted to find out. I was hooked.

Those three months on the surgery rotation ended up having a huge impact on my life. In many ways, I stumbled into surgery. Still, it felt like a calling—albeit, one that involved a massive commitment and an overwhelming responsibility: to take someone's life in your hands and take responsibility for what happened next. The satisfaction of doing an operation well, of pushing yourself to get everything right and make someone better—it is intoxicating, almost godlike. I also loved the team I worked with during those three months; we functioned like a well-oiled machine.

Toward the end of our third year, our advisors asked us what field we wanted to go into. I couldn't stop thinking about that kidney transplant. I had scrubbed in on a few more during my time on the surgery rotation, and I remained fascinated. I never saw a liver or a pancreas or a heart transplant. Still, I could not get over that we could take a kidney from someone who had just died and plug it into someone else, maybe even a day later, and it would start working. It seemed so simple. I wondered then if I would ever be able to do something like that myself.

WHILE ALEXIS CARREL showed it might be technically possible to transplant a kidney from one living being to another, this alone did little to alleviate the desperation both of patients with renal failure, and of their doctors, in the first half of the twentieth century. It would take another advance in the care of these patients before transplant could be considered a clinical option. Enter the remarkable story of dialysis.

Something about dialysis units always reminds me of the film *The Matrix*—not the part where Neo is able to dodge bullets while moving in slow motion, but the part where people are plugged into the matrix through sockets at the back of their necks. That is more or less how dialysis works, and it seems totally barbaric.

When someone's kidneys stop working, the blood has to be filtered in some way, or the patient will die. The most common method of dialysis is hemodialysis, or dialysis through the blood. In order to perform hemodialysis, we first establish an access where the needle can be inserted. Typically, a surgeon will make an incision on either your forearm or your upper arm and sew the end of a vein onto an artery (often either at your wrist or your elbow). The vein will swell up as it is filled with the pressure of arterialized blood, becoming big like a sausage. Once this connection heals, and the vein wall thickens, big needles can be plunged into this vessel at the arterial and venous sides, with blood being pulled out from the needle near the artery so that it can flow through a dialysis machine. Here, it runs through a membrane surrounded by a bath that pulls off electrolytes and toxins, and then returns through the needle poked in the venous side. This process takes three or four hours, and the whole time, the patient sits there with his arm out straight so the tubing doesn't kink and the needles don't come out. We try to put the socket (actually called a fistula) in the nondominant arm, so the patient can use his or her other hand to write or read.

It is truly a miserable process. In addition to being confined to a chair for four hours three times a week, many patients feel lousy during and after the sessions, with symptoms that include fatigue, coldness, headache, and muscle cramps. After dialysis, patients will often spend the rest of that day lying in bed. As many patients have described it to me, dialysis keeps you alive but is no way to live. But then again, what's the alternative?

When dialysis was first introduced, the man responsible would never in a million years have considered it an option for long-term treatment. At the time, Willem Kolff was building his dialysis machine in secret in Nazi-occupied Holland, using sausage casings and the motor of a sewing machine, all the while helping the Dutch resistance against the Germans.

Kolff was born in Leiden, in the Netherlands, in 1911. His father, who started as a family physician, ultimately settled on running a tuberculosis sanatorium. Kolff had significant exposure to his father's practice, and he became fascinated with medicine. He also had a knack for woodworking and mechanical fixes. What he enjoyed most about the prospect of a career in medicine was the opportunity to solve problems, particularly by building something with his hands.

Kolff finished medical school in 1938 and began his practice. There are numerous examples of him inventing things to help his patients, including an early version of a sequential-compression device— "squeezy boots," as we call them, which strap around patients' legs and periodically inflate and deflate to prevent blood clots. Kolff's first real exposure to renal failure and the helplessness associated with caring for patients with this disease was with a patient named Jan Bruning, twenty-seven, who died before his eyes. Bruning had Bright's disease, a historical term for many different causes of kidney disease, some of which can recover and some of which cannot. Back in the 1930s, there were all kinds of useless treatments for Bright's, including dietary changes, bloodletting (rarely a good option), and baths (which sounds nice, at least). Yet this didn't sit right with Kolff. He hadn't become a doctor to watch some young man die in front of him while giving him a bath. He figured it this way: kidneys are just filters that clean waste products from the blood, and he knew one of the key toxins was urea. How hard could it be, then, to filter the blood? If he could just clean out the blood for a short period, perhaps he could give the kidneys some time to rest and recover.

This idea had been tried a couple of times prior to World War I, but with little success. At that time, the type of membrane that would allow molecules of certain sizes to flow across a gradient, but that would be impermeable to bigger molecules, including blood cells and important proteins, didn't yet exist. In addition, any blood placed in a container for exposure to such a membrane would clot.

Kolff knew he could get over both these barriers because . . . he liked
sausages—or, at least, he was aware that sausages were encased in
cellophane, an edible artificial membrane made from regenerated
cellulose that allows sausages to keep their shape and allows flavors
to diffuse across the membrane. He also was aware that cellophane
was already being used as a filter for purifying fruit juice. It seemed
to Kolff and a few of his colleagues that if he could expose blood to a
cellophane barrier with a lot of surface area, and have, on the other
side of the membrane, a fluid bath without urea (and without any
other protein or electrolyte he wanted to remove), he could cleanse
(or dialyze) the blood. He drew off enough of his own blood to fill a
sausage casing and mixed this with urea at a concentration he estimated would be found in the blood of a patient in renal failure. He
then poured this into the casing and placed it on a board floating in
a bath of water. He added a small motor to the board, so it would
rock back and forth, sloshing the blood around and thus allowing
it to come in contact with the cellophane membrane. Five minutes
later, he retrieved the blood sample and, to his surprise, found that
almost all the urea had been transferred over to the bath! Thus,
dialysis was invented.

There were still challenges, though: having enough surface area
for blood to contact the cellophane, avoiding clots while the blood
was being filtered, figuring out how to get the blood from the patient into the filter system and back, not to mention knowing how
much blood to pull out and then put back in without killing the
patient. Yet, as with so many pioneers, the right combination of
intelligence, vision, conviction, an ability to ignore the naysayers,
and near obsession regardless of these barriers made Kolff the man
for the job. Still, in addition to the technical challenges, Kolff had
to deal with one other major impediment: the Nazis.

The Netherlands was invaded on May 10, 1940, and the Dutch army defeated in less than a week. This was devastating for Kolff, who despised Nazi policies. He had many Jewish friends and colleagues and had witnessed their deportation, murder, and, sadly, many suicides among them. Perhaps most impactful for him had been the death of his boss and mentor, Leonard Polak Daniels, one of the few established physicians who believed in Kolff's visionary medical inventions. After the Germans invaded and conquered the Netherlands, Daniels killed himself rather than be taken by Hitler's army. His replacement at the large university hospital in Groningen was a staunch supporter of Hitler, a fact that prompted Kolff to secure an appointment in a small hospital in the town of Kampen, effective in July 1941. This move proved to be fortuitous, as it gave Kolff the chance to move forward with his plans without much scrutiny.

Kolff had two main goals when he went to Kampen. The first was to treat as many patients as possible, including those with renal failure. The second was to save as many people as possible from the Germans. To prevent the deportation of suspected members of the resistance, he made up fake illnesses for them, shielded others who were being investigated by hiring them, became involved in an aborted assassination plot to kill the Nazi head of police—he was to have driven the getaway car—and even agreed (though was not in the end required) to cut up and dispose of the body of a Jewish woman who had died while friends were hiding her from the Nazis. He also administered medicine to people that made their skin turn yellow, so the Nazis would think they had jaundice and couldn't work in their camps.

Amid all this, Kolff also found time to build the first dialysis machine. After the success of his original experiment using his own

blood, he let his mind wander regarding how to increase the exposure of blood to the cellophane and how to get blood to flow out
of the patient, through the cellophane casing, and back into the
patient. In 1942, he thought he had the answer. Early one morning,
he walked over to an enamel works owned by Hendrik Berk, and
together, Kolff, Berk, and Berk's engineer E. C. van Dijk came up
with a plan for a machine.

It really was quite simple. Blood was drawn from a patient and
flowed into a central shaft within a large cylinder. The shaft had multiple spokes that connected out to cellophane tubing. This tubing,
which was long and thin, wrapped in spirals around the big cylinder.
The cylinder was suspended horizontally in the dialysate (the fluid
that would serve to pull the toxins out of the blood) and attached to a
motor that allowed it to spin. With each turn, the blood, obeying the
laws of gravity, flowed into the cellophane, which became immersed
in the dialysate as the cylinder spun. The toxins and electrolytes
flowed across the permeable membrane and into the dialysate bath
(which was made up of a low concentration of sodium chloride, sodium bicarbonate, and potassium chloride mixed in a large quantity
of tap water. Kolff was constantly monitoring electrolytes and urea in
the patients' blood, and would make adjustments to the bath depending on these levels and how fast he wanted to correct imbalances.
Eventually he added glucose to the bath as well, to help pull water
from the patients' blood and prevent disequilibrium of electrolytes.
Although Kolff always described dialysis as simple, it is actually quite
complex, and his constant attention and analysis during sessions was
as important as any of his innovations in the ultimate success of his
machines). It was a closed circuit that allowed the blood to progress
from the patient through the rotating loops of cellophane while being
exposed to the dialysate, and then back into the patient.

The first dialysis machine had been built. Now all Kolff had to do was see if it worked. For this, he went straight to human patients, choosing those who clearly were going to die without any intervention. His first attempt was less than stellar. He tried the machine on an elderly Jewish patient who was so ill the Germans didn't bother to deport him to a concentration camp with the rest of his family. Kolff initially had trouble getting blood out of the man's brittle arteries, and in the end was able to inject a mere fifty milliliters of blood through the machine. Then the cellophane sprang a leak, causing the bath to become foamy red and spill all over the floor. His second patient was a much better candidate—a twenty-eight-year-old woman who had been previously healthy but who, for unknown reasons, had developed kidney failure, presenting with high blood pressure, confusion, loss of vision, and palpitations. She was found to be in renal failure with an extremely high level of urea in her blood. Kolff thought there was a chance that if he cleansed her blood for a few days, her kidneys could recover. For the first session, he removed half a liter of blood from an artery in the patient's wrist, ran it through his machine, and returned it to her through a needle in a vein in her arm. She regained consciousness and seemed better. Kolff watched her for a day, and when nothing bad happened, he decided to resume dialysis. By this time, he had designed a pulley system whereby he could lower portions of the contraption to allow blood to come into the machine and raise different parts when he wanted the blood to return to the patient. The patient underwent twelve sessions of dialysis in all, the tenth session lasting six hours. Throughout the process, Kolff carefully followed the woman's lab results, including electrolyte and urea levels, and everything was correcting nicely. By the end, he ran twenty liters of blood through his machine during a single session, as much as four times the patient's blood volume. Her blood pressure normalized, and her mental status had improved. By the end, Kolff was hooking her directly to the machine and letting her blood flow through the dialyzer and back into her body—truly the first case of continuous dialysis ever performed. This was 1943. By the twenty-sixth day, Kolff finally had to stop. The dialyzer was still working, but the woman's kidneys had not recovered, and he couldn't find any more blood vessels. The needles Kolff had access to were very primitive, and each vessel could only be punctured once. With every new session he had to find a new artery and vein to access. She died shortly thereafter from renal failure.

Kolff immediately got to work on building a second, even bigger machine, this time constructed out of wood; with the war raging, aluminum was no longer available in the Netherlands. He moved his first machine over to a hospital in The Hague and set up his new one in Kampen. Then he got the word out that he was looking for patients. He even tried to hold weekly conferences to discuss his capabilities. Somehow, despite the worsening conditions in the war, he managed to build a third machine, which he placed in Amsterdam.

Over a two-year period, Willem Kolff dialyzed sixteen patients secretly at night. Only one patient survived, but Kolff was the first to admit that it wasn't from his dialysis. Even so, he knew he had made great progress. He had made many improvements to his machines, and was now able to achieve a flow rate of 150 milliliters of blood per minute through his tubing, which was, all told, about 45 meters long. He was convinced that if he could just get the right patient, someone who wasn't so far along in his kidney disease that it was too late and whose native kidney function could recover, his machine would work.

He finally got his chance in 1945, when peace returned to the

Netherlands. Ironically, that first successful patient would be an imprisoned Nazi sympathizer. Her name was Sofia Schafstadt, a sixty-seven-year-old with an inflamed gall bladder. Her illness and the antibiotics she was taking to treat her infected gall bladder had caused her kidneys to fail. Over an eight-day period she had made almost no urine, her urea levels were dangerously high, and she was slipping in and out of a coma. Kolff finally convinced her team to let him dialyze her, and because they figured she was going to die anyway (and because she was a Nazi sympathizer and they didn't really care), they let him.

The first session lasted more than eleven hours, and by the end the patient was awake, the urea level in her blood had normalized, and her blood pressure had come down to a safe level. Kolff watched her closely over the next day, and when he was getting ready to hook her up to the machine again, she started to make urine on her own. Kolff was sure that without the dialysis session, she would have died.

How remarkable that Kolff was able to accomplish all that he did in the environment in which he worked. But he didn't stop there. After the war, he and his assistants traveled the world, telling anyone who would listen about his invention. He gave away his beloved dialysis machines and, once he ran out of the machines, provided blueprints for constructing them. It probably never occurred to him that dialysis would be used as a chronic treatment for kidney disease. I imagine he would have been as shocked as I was the first time I walked into a dialysis unit as a medical student and saw countless people sitting in lounge chairs, large tubing filled with blood running from needles in their extended arms into mysterious whirring machines that would periodically emit shrill alarms that everyone but I seemed to ignore. The machines look so complicated and industrial now that I had never realized how simple the concept and early design for them was until I started researching this book.

Kolff saw dialysis as a temporary measure, something that would give the patients' native kidneys time to recover. When it became clear that the majority of his patients would not see such a recovery, he turned his attention to the next step on the road to curing them: kidney transplantation. He ultimately made his way to the Cleveland Clinic, where he became involved in its kidney transplant program, and he remained active in the transplant community throughout his career. He was also one of the innovators in the development of the membrane oxygenator for cardiac bypass, and ended up at the University of Utah, where he was one of the inventors of the most famous version of the artificial heart. Kolff's contributions to the field of organ replacement are truly legendary, and his persistence in establishing dialysis allowed others to take the next step in achieving success in organ transplantation. It was the advent of dialysis that allowed a few premier hospitals to become destination centers for patients in renal failure, and that bought their physicians time to think about more permanent ways to treat their disease. Yet before this could be a possibility, before someone could build on the advances of Alexis Carrel and Willem Kolff, someone had to crack the barrier of the immune system.

| 4 |

Skin Harvest

I cannot give any scientist of any age better advice than this: the intensity of the conviction that a hypothesis is true has no bearing on whether it is true or not.

—PETER MEDAWAR, *ADVICE TO A YOUNG SCIENTIST*

The first time I ever took from death was on a crisp October evening during my second year of medical school and a full year before I witnessed my first kidney transplant. Of all the ghoulish things I have been involved with during my career as a transplant surgeon, what happened that evening has to be the most bizarre.

I had just started working for the New York Firefighters Skin Bank, an organization set up by the burn center at New York Hospital in 1978 to recover and store skin from recently deceased donors. A couple of medical students from each class were handpicked to work for the skin bank, to join an "elite" group that would head out in the middle of the night to skin dead people. Of course, there was a purpose to it: the skin would be used as temporary grafts for burn victims, to give them coverage while they healed enough to undergo grafts with skin from their own bodies.

Picture the case of a factory worker who falls backward into a vat of boiling oil, or a young man who gets caught in a house fire when his meth lab explodes—these are two actual patients I took care of years later, in my residency. Both these men came in with more than 80 percent of their bodies covered in full-thickness burns. With the loss of that much skin, our barrier to the outside world, both men were losing fluids and electrolytes through their open wounds, couldn't maintain their body temperatures, and were at high risk of infection from the bacteria we encounter every day. In addition to resuscitating them with aggressive IV fluid infusion, it is critical to get some sort of coverage over their wounds, ideally by taking skin from preserved (that is, unburned) sites on their own bodies, so that the grafts take and are not rejected. But given the extensive nature of these two patients' burns, it would take months and countless trips to the operating room to harvest enough skin to cover the open areas. So, in their case, we had to use someone else's skin.

We know that using a graft from someone else will surely end in rejection, but we also know that patients who have suffered serious burns generally have a poor immune response—which, ironically, allows a skin graft from a donor to last for weeks, much longer than it would in a healthy recipient. And this temporary coverage can buy enough time for these very ill patients to be stabilized.

Nowadays, skin substitutes compete with cadaveric skin, but back in the 1990s, when I was working at the skin bank, no other substitute had been approved, so donors had to be used.

At the time I was chosen to harvest donor skin, I didn't have any concept of who had figured out how to do it, how skin grafts related to the field of organ transplantation, or how relevant this work would eventually become to my life. As a second-year medical student, I hadn't yet started my clinical rotations where we see patients, had never actually taken care of a patient with an illness, and still had it in my head that I would become a pediatric oncologist. I just thought it would be fun to be part of this team and was intrigued by the idea of learning new skills and spending some time in an operating room.

Somehow, I had managed to get myself chosen for this skin gig, had gone through the training, and was now going on my first "run." Accompanying me were Brian and Lawrence, two more seasoned members of the team. Brian, who had been a part of the skin bank for about a year, was a fellow medical student, and would ultimately become a close friend. Lawrence, the most experienced "banker" around, was a graduate student and a sight to be seen. He was at least six and a half feet tall, with long blond hair, built like a truck, with a personality to match. To prepare for the run, I went up to the lab tucked away on the twenty-third floor of New York Hospital and loaded up a cart with the necessary supplies: sterile drapes, gloves, gowns, sponges, and any number of disposable items one sees in an operating room. The most memorable pieces of equipment were the infusion pump and the dermatome, with its blade—a kind of lawn mower for skin.

As the new guy, I got to wheel this huge cart down to the van, a big red vehicle with "New York Firefighters Skin Bank" stenciled on the side in gold letters alongside a picture of a fireman pulling a child from a burning house. The donor was somewhere out on Long Island, and we took the Queensboro Bridge that night. I sat mostly in silence, going over the steps of the procurement in my head.

When we got to the hospital, we were told that the donor was still in the operating room, where the other procurement teams had just finished with him. I would later learn that when the donors aren't donating other organs, we have to pick them up in the morgue—an experience I would eventually come to find macabre. I put on my mask, shoe covers, and hat and wheeled the equipment cart into the OR. And there he was. My first recently dead patient.

There was no denying that he had just been alive. You could see it in his face; in the stubble on his jaw, which hadn't been shaved in a couple of days; in his eyes, his hands. At the sight of him, I pictured my own dad (who was, and is, still alive and healthy), thinking that he might have looked like that in death. As I was drifting off into these thoughts, starting to wonder what the hell we were about to do, a piercing voice pulled me back in.

"Let's go, dumbass!"

The procurement of organs is a surreal process. (In those days, we still called it harvesting, a term we've since dropped in favor of the more respectful "procurement.") What I remember is seeing the long incision running from the top of the donor's chest all the way to his pubis, an incision that had been stitched back up with a big suture like you'd see on a baseball, only black, indicating that the chest and abdominal teams had preceded us. The other thing I remember were the long incisions running down both legs, made by the bone guys, who would have replaced the bones they'd taken with broom handles, so that there would be some structure to the legs when we (or the funeral parlor) moved the body around. The donor's eyes were taped shut, signaling that the eye team must also have slipped in before us.

We flipped the donor onto his front and prepped him, just as I would do years later on living people. Then we scrubbed in and draped him with sterile sheets. Next came the weird part.

Brian turned on the infusion pump, which would allow us to

pump saline, now in two large bags hanging high above on IV poles, through a couple of long, sterile plastic extension tubes and into the donor's body through sixteen-gauge needles attached to the ends of the tubes. The pump whirled into action, generating a soothing, rhythmic sound, and with saline squirting out of the needles and into the air, Lawrence began poking the needles into the donor's back, and his skin started to lift up like a balloon, turning this guy into the Stay Puft Marshmallow Man. Once we were done inflating the donor's skin, we opened a few bottles of mineral oil and greased his back and both his legs. Then I watched as, with one stroke of the dermatome, Lawrence removed a piece of skin from the top of the donor's back to just above the ankle. (I remember thinking how incongruous it was to see such a graceful move from such a big, scary guy as Lawrence.)

We finished skinning the donor's back, taking turns getting a feel for the dermatome, which resembles an electric paint remover and works like a razor. It is fitted with a long sharp blade, and you set the distance from the blade to the dermatome surface, depending on the width of skin you want to harvest. For a perfect strip running from the back to the ankle, you have to constantly adjust the width and the pressure you apply as the thickness of the skin changes. Lawrence was a master at this. (By the end, I can happily say I managed to harvest a few nice strips.) We then flipped him over onto his back and repeated the process, harvesting skin from his chest and the front of his legs.

Many more nights like this would follow, and I finally mastered the art of harvesting skin. There was always a small moment, during each case, when I would think about who the donor might have been in life—but it was fleeting. I had a job to do, and found it easier to focus on that. In a typical case, we'd have music blaring, tell jokes, talk about things that seemed important in our lives, where we would go to eat after we were done (a tradition after every skin run)—never really grasping, I confess, how incompatible these thoughts might have been with the tasks we were performing.

In a few cases, we arrived early (or the teams before us were late), and I would get to scrub in with the organ transplant teams. They would have flown in from all over the country, coming to get the heart, lungs, liver, kidneys, while back home, their recipients waited to find out if tonight was the night their lives would be saved. I remember thinking how cool it would have been to fly back with those guys to their respective hospitals and watch the organs they were harvesting come back to life. I imagined it as an adventure.

At the time, the skin seemed so meaningless compared to hearts and kidneys and livers. But then, as I advanced in my career, I slowly came to understand that the gift of any organ, be it a liver, kidney, heart—or, for that matter, bones, eyes, heart valves, and, yes, skin—is truly a wonderful gift. Not to mention that skin is the tissue that cracked the code to organ transplantation. Indeed, without skin, and Peter Medawar, there would be no transplant surgery.

North Oxford, England, The Blitz, 1940

Peter Medawar was enjoying a Sunday afternoon with his wife and daughter in the garden of their house in Oxford when they saw a two-engine plane approaching in the sky. Expecting this to be an attacking German bomber, Dr. Medawar and his wife grabbed their daughter and rushed into their air-raid shelter, a structure that had become so common in English homes after the start of World War II. They heard a loud explosion two hundred yards away. It turned out this wasn't a German plane making a bombing run, but an English plane in distress.

An airman survived the crash and was brought to the local Radcliffe Infirmary with third-degree burns all over his body. Knowing that caring for him would be almost futile, and with his chance of survival next to nothing, his physicians came to Dr. Medawar to ask for help. Was he a celebrated trauma surgeon? A critical-care specialist with years of success attending to the sickest patients? No, he was a twenty-five-year-old zoologist whose most significant work had been in the discipline of cell culture, studying the mathematical basis of growth in . . . embryonic chicken hearts. Was Medawar familiar with the work of Alexis Carrel spanning the previous half century? Did he know that Carrel had successfully transplanted organs only to watch them stop working over a few days from some mystical "reaction"? If he did, he certainly wasn't focusing his intellectual energies on these things. And he certainly didn't know about the efforts of Willem Kolff, just 350 miles away.

Medawar was born in 1915 in the city of Rio de Janeiro, Brazil, to an English mother and a Lebanese father who worked for a dental supply manufacturer. He moved to England as a small child at the end of the First World War and stayed there for his education while his parents returned to Brazil. He was able to make it through trying years in various boarding schools around England, and ultimately entered Oxford University's Magdalen College in 1932.

To be fair, when the physicians taking care of that young English pilot came to Medawar for help, he wasn't a complete neophyte in the study of burns. After World War II began, the Recruiting Board told him he had a responsibility to do research that might help the war effort. So he began using his tissue culture system to investigate which antibiotics would be effective and nontoxic to burn wounds, known to be at high risk for infection. He published papers touting sulfadiazine and penicillin, an important finding in those days, but nothing compared to what was coming next. Then, some combination of the horrors of life in England in 1940 and good mentorship turned his attention to the study of burns. And that changed everything.

For many years, I couldn't understand why Sir Peter Medawar was considered the father of transplantation. Mcdawar's most famous discovery was the concept of "acquired immunological tolerance." He found that if he injected a fetal mouse from one genetic family (directly through the pregnant mother) with cells of an immunologically mismatched donor (meaning a different mouse that wasn't genetically identical), the recipient mouse would then accept skin grafts from this same donor type without rejection once it grew into an adult, with no need for any medications to block an immune response. In other words, it was "tolerant" to that donor. He presented his initial findings on this topic at a conference in 1944, and ultimately published a more complete report in 1953. This idea of tolerance, which some people have called the "holy grail" of transplantation, is not a state we achieve or try to achieve in modern transplantation, except in animal research or some very small experimental protocols. Instead we treat patients with chronic immunosuppression to prevent rejection of their organs.

Prior to Medawar, all attempts at transplanting human organs—and there were many—ended in complete failure. Organs would be sewn in and quickly die (along with the recipients), and no one knew why. Carrel, at the turn of the century, thought there was some sort of "biological force" that prevented successful transplantation of organs. The idea of an immune response was totally foreign in those days. Most level-headed people had given up on organ transplantation as anything more than a crackpot idea that some crazy scientists were doing in the lab.

If Alexis Carrel displayed the persistence and physical genius that was necessary for the technique of transplanting an organ from one animal to another, then Peter Medawar took the next step and demonstrated that it might be possible to overcome this "biological force" and have the transplant enjoy sustained function. Medawar brought credibility to the field, providing researchers with a legitimate mechanism to study, and a vision that would inspire so many to make transplant a reality.

He began by trying to solve the problem facing the burned pilot. First, he addressed the question of how to expand the small amount of skin left to sufficiently cover the remaining 60 percent of the pilot's body. He initially approached this dilemma with tissue culture, obtaining skin left over from plastic surgery operations and trying to get the cells to grow. No luck. Next, he tried to take the autologous skin (that is, the pilot's own skin) and slice it thinner and thinner, in an attempt to maximize burn coverage from what was available. That didn't work, either, and the pilot ended up dying.

Medawar's frustration led him to believe he should explore the use of homografts, grafts from donors (now called allografts—both terms refer to grafts from the same species as the recipient) as opposed to autologous grafts. He successfully wrote a grant to the British government to study this and went to work in the burn unit of the Glasgow Royal Infirmary. The first thing he did, with his collaborator, Tom Gibson (a surgeon), was to experiment on an epileptic who had suffered severe burns after falling into a gas fire. With the help of Gibson, Medawar placed numerous small homografts four to six millimeters across, on the woman's burns, next to autografts, taken from her own unburned skin, as controls. They got the homografts from willing volunteers (probably medical students). Grafts were removed at regular intervals and examined under a microscope. Medawar noticed that the homografts were invaded by lymphocytes, the white blood cells of the immune system, whereas the autografts were accepted on the recipient, with ingrowth of blood vessels (meaning recipient blood vessels grew into the donor graft) and minimal inflammation. Next, Medawar and Gibson placed a second set of skin grafts from the same donors as the first, to see if they would last as long as the first set. They found this second set was destroyed almost immediately, with an even stronger inflammatory reaction. Medawar published these results in an article entitled "The Fate of Skin Homografts in Man."

What made Medawar so impactful was his ability to be persistent; admit to mistakes; follow a course of experiments over months and years, in order to tell a complete story; be right about most of the things he did; present his data at international conferences; mentor students and junior colleagues who went on to do great things themselves; and most important, publish.

After returning to Oxford, Medawar turned his full attention to testing the hypothesis that homograft rejection was an immunologic phenomenon. He knew he couldn't study this in great detail in humans, so he learned to do skin transplantation in rabbits, mice, guinea pigs, and cattle. He was joined by his first graduate student, Rupert Everett Billingham, who played a huge role. Then an encounter changed everything.

Medawar was at the International Congress of Genetics in Stockholm when he met a friendly Kiwi by the name of Dr. Hugh Donald. They got into a conversation about distinguishing between fraternal and identical cattle twins. Donald was trying to identify characteristics that were based on genetic differences versus the environment, but he couldn't figure out an easy way to distinguish between identical versus fraternal twin calves shortly after birth. Medawar thought that would be easy.

"'My dear fellow,' I said in the rather spacious and expansive way that one is tempted to adopt at international congresses, 'in principle the solution is extremely easy: just exchange skin grafts between the twins and see how long they last. If they last indefinitely you can be sure these are identical twins, but if they are thrown off after a week or two you can classify them with equal certainty as fraternal twins.'"

It turns out Donald kept his cattle a mere forty miles from Birmingham, where Medawar was working at the time, so he invited Medawar out to do the skin grafts. Medawar and Billingham had no interest in going out to a farm, but true to their word, they accepted the invitation. Lo and behold, all the grafts were accepted!

Medawar questioned his hypothesis rather than the data. He delved into the literature to try to understand what he was missing and found the answer in Madison, Wisconsin—cow country, of course.

Immunogenetics Laboratory, University of Wisconsin, 1944

Ray Owen was working as a postdoctoral fellow in the lab of L. J. Cole the day the letter from Maryland was delivered. It described a pair of twin calves that appeared to have different fathers. The cattle breeder had mated his Guernsey cow with a Guernsey bull. The Guernsey ended up delivering twins, but it was clear from the color patterns that the twins had different fathers. Owen was fascinated by this story, and asked to be sent some blood. He found that the calves had identical blood groups, despite the fact that they weren't identical twins—they were of different sexes—and had different fathers. He further identified that each twin had blood group antigens from the mother and from *both* fathers circulating through its bloodstream. Each twin had two blood types, a finding that had never been described before! How could that be?

It was known at that time that cattle twins in utero, unlike humans, share blood vessel connections, and hence can exchange blood while embryos. It was even known that these connections were the reason "freemartin" cattle—that is, female calves who are twins to a male—were sterile. (Hormones of the male twin suppress sexual development in the female, a concept that was first described in 1916.) But even though the blood would have been shared in utero, the red blood cells that had come from the twin would have been expected to die off after birth, leaving the calves with a single blood type. But the idea that these cells remained present for life was startling. This implied that blood precursors were being shared, not just the red cells themselves. These twins were chimeric—that is, they harbored cells throughout their lifetime derived from the genes of two different fathers.

Owen published the findings on red cell chimerism in *Science* in 1945. In the version he submitted, he discussed the concept of immune tolerance, and how this could be applied to organ transplantation someday. Sadly, the reviewers at *Science* thought this was more science fiction than science and rejected this portion of the paper.

Back to England, 1949

Upon reading Owen's original paper, Medawar and Billingham suddenly understood what had happened. The nonidentical twin calves had accepted the skin transplants because they had been exposed to each other's cells during their development in utero, and were therefore chimeric, probably not just in regard to their red blood cells, but also other cells of the immune system. The researchers quickly published their findings and moved on to the next set of experiments: to find a strain combination of mice in which fetal tolerance worked. They succeeded in making mice tolerant to skin grafts from unrelated mice by injecting donor cells in the fetuses of the recipient mice in utero. In other words, they found one technique to get over the insurmountable barrier of transplant rejection and named the phenomenon "acquired immunological tolerance." Medawar and Billingham published these findings in *Nature* in 1953, along with their graduate student Leslie Brent. As Medawar said:

> *The real significance of the discovery of immunological tolerance was to show that the problem of transplanting tissues from one individual to another was soluble, even though the experimental methods we had developed in the laboratory could not be applied to human beings. What had been established for the first time was the possibility of breaking down the natural barrier that prohibits the transplantation of genetically foreign tissues: some people had maintained that this was in principle impossible . . . Thus the ultimate importance of the discovery of tolerance turned out to be not practical, but moral. It put new heart into the many biologists and surgeons who were working to make it possible to graft, for example, kidneys from one person to another.*

This was the first time anyone had ever been able to transplant tissue between any living creatures and have the graft survive.

Of course, I didn't know anything about Medawar as I drove around New York in the middle of the night skinning dead people. But when I did get the chance to scrub in with those organ procurement teams—and watched as the organs were wheeled out in their separate coolers to be flown off into the night, temporarily sleeping before they would be filled again with the warm blood of a new owner and jump back to life as if nothing had happened—I wondered how anyone could possibly have thought this would work. It would never have crossed my mind that it all started with a British zoologist doing skin grafts on mice.

The effect Peter Medawar's finding had on the small numbers of surgeons and researchers who were dabbling in the field cannot be overstated. The first three pieces of the puzzle were in place: the technical proof that an organ could be taken from one animal and sewn into another and start working, demonstrated by Carrel; Kolff's mechanical method of keeping patients in renal failure alive long enough to develop realistic strategies for transplantation in humans; and Medawar's immunologic proof that there could be strategies to overcome the "biological force" leading to rejection of these organs. With his elegance, honesty, and optimism, Medawar gave hope to those who would follow him that clinical transplantation could become a reality, and inspired a whole generation of investigators to dive headfirst into the game.

| 5 |

Kidney Beans

Making Kidney Transplant a Reality

I have not failed. I've just found 10,000 ways that won't work.

—THOMAS A. EDISON

Success is not final, failure is not fatal: it is the courage to continue that counts.

—WINSTON CHURCHILL

My path from that operating room on Long Island for my first skin harvest to an operating room in Madison, Wisconsin, transplanting my first kidney as an attending surgeon was anything but straight and easy. It spanned more than thirteen years, four cities, and many all-nighters taking care of patients and needy attending surgeons. In that time, I flirted with the idea of becoming a heart surgeon, a laparoscopic surgeon, and a barista—sometimes I still wish I had— but the excitement of watching a donated kidney make urine in the recipient's body never left my mind. Throughout my early training, I still wondered whether I could ever actually be the one responsible for making the gift of a transplant happen. Yet as my training progressed, I started to think I was ready. I had scrubbed in on hundreds of kidney transplants during my fellowship, performed all the operations (at least in my mind), taken care of the patients day and night, week after week, and always seemed to know what to do. But of course, there always was an attending present who was ultimately responsible and whom I could silently blame if something went wrong.

All that changed on my first day as an attending surgeon. Suddenly, almost any decision seemed too difficult. For the first time in my life, no one else would have to double-check my judgment or sign off on my assessment. I could take a patient to the operating room, cut him up, and pull something out, and no one would ask me if I was sure, what my reasoning was, and if I had considered other options. It was terrifying!

My first kidney transplant was between a wife, a living donor, and her husband, the recipient. He was a redo; it was his second transplant. I met the couple a few days before the surgery. The wife was a nurse, which added to my stress—I was sure she could smell my fear and lack of experience. I examined the recipient. He was a pretty big guy, and I thought the surgery was going to be tough. (It's always harder to operate on bigger, obese patients. The vessels are deeper and surrounded by fat, making the operation take place in a deep hole. It has always amazed me that people outside of surgery are surprised by this; I suppose they want to believe that all surgeries are the same and predictable.) I started talking to them about the operation, the risks, the recovery, and at some point, the wife interrupted me and asked, "So, how many of these have you done yourself?"

I wondered if something in my shaking voice had given me away. I wasn't sure of the best way to handle this and finally said, "Well, this is my first as staff, but I did a couple hundred as a fellow and feel very comfortable with it."

I hoped they would ask for a different surgeon. They were both quiet for a minute, and then the husband said, "I think God wants you to do it."

I felt like saying, "No, trust me. God does *not* want me to do it."

Thankfully, the recipient's wife had kept her wits about her. "I think you should get someone else. Nothing personal to you."

But the husband said, "No. I really think I want him to do it."

"Okay, great," I stammered. "I'll take good care of you."

"Damn, why is this so freaking tough?" I later said to my fellow as we sewed the vein in. I had opened the recipient on the left side and exposed his iliac artery and vein. The vein was so deep, his abdomen so thick, and the space so small that I could barely see anything. I probably should have made a bigger incision, done something different with my retractors, but at this point I wasn't going to turn back. I was sweating profusely, my hands were shaking, and I even had them turn down the music in the OR. Where was God?

Once the vein was done, which took twice as long as I thought it would, we moved on to the artery, and I managed to sew it in. Now came the moment of truth.

"Reperfusion," I told anesthesia. "Give my creation life!" I released the clamp on the vein, and then the two clamps on the artery, distal and proximal, and then stared at the pale, flaccid kidney, waiting for it to turn pink and firm.

Nothing. Nada.

I felt the artery. No pulse. My heart started to race. Had I "back-walled" the artery—meaning, had I caught the back wall of it while sewing the front? Or had I raised a flap of the intima (the inner layer of the artery, which has three layers), causing a dissection that was now blocking flow into the kidney? Should I take the donor kidney out, flush it, and repeat the whole thing?

Then I remembered the retractors. Sometimes when they are in deep (as they were with this recipient), they compress the iliac vessels, preventing blood from flowing through them and into the kidney. I pulled the retractors partway up and out of the wound and—voilà! The kidney filled with blood and turned a beautiful pink. Shortly thereafter, urine started to squirt out of the ureter. I was overwhelmed by a feeling of joyful satisfaction.

Since then, I have done hundreds of kidney transplants, and I promise much more smoothly than that first one. To this day, though, I experience the same feeling of amazement when the organ pinks up and urine squirts out. To this day, I still can't believe it works—and not just for a few days or a few months. With a little luck, the little beans I successfully transplant into patients should keep pumping out urine for years.

How did we get to this point, where surgeons can take a kidney out of someone, alive or dead, and successfully put it into someone else? Was it always this easy?

Boston, Massachusetts, Just Before Midnight,
Unspecified Date in 1947

How must David Hume have felt as he ran to the inpatient ward holding a most unusual package in a sterile pan, hoping no one would see him? The fact that it was midnight, that he was bringing the item to a back room with almost no permanent lighting (rather than the operating room), suggests that he knew very well that what he was about to take part in was not entirely kosher. The patient was a twenty-nine-year-old housewife who had been admitted to the then-named Peter Bent Brigham Hospital after undergoing an illegal, unsterile abortion in her fourth or fifth month of pregnancy. The procedure had led to sepsis, blood hemolysis, and ultimately renal failure. She was treated with antibiotics, and overall, she responded, but she was making no urine. It became obvious to everyone involved that she would soon die. If this were to happen today, she would be placed on temporary dialysis until her kidneys recovered. But in 1947 in Boston, dialysis was not yet available. So, what to do?

Hume and another surgeon, Charles Hufnagel, had been on the lookout for a possible kidney donor. They had both done numerous experiments in dogs by this point, and they knew that a transplanted kidney would last a week at most before it ceased to function. Perhaps a few days of function would give this young woman's kidneys enough time to kick in. What did they have to lose?

On this fateful night, there was a stroke of luck. A surgical patient at the hospital died during surgery. A relative of the deceased who happened to be a hospital employee agreed to let Hume take one of the now-useless kidneys to save the life of the young woman. Hume opened the dead donor and removed one of the kidneys carefully and placed it in a basin. How much time had passed from death to procurement is unknown. Without a doubt, enough time had passed that the donor was truly a corpse, not a warm, fresh donor we are used to nowadays.

Hume and his team would have liked to sew this kidney in place in an OR, where they would have had good lighting, sterile equipment, nursing assistance, and the space to work. But apparently there was "administrative objection to bringing the patient to the operating room." Whether this was because of the critical nature of her illness, the illegal abortion she'd undergone, the low likelihood that this surgery would work, or the general resistance to desecrating a dead patient for such a desperate and futile experiment, the decision was made to do the surgery in a side room off the inpatient ward.

Two gooseneck lamps were pulled into the room. The woman's arm was placed on a table and prepared sterilely with alcohol. An incision was made in the anterior surface of her elbow. Hume and Hufnagel dissected out the brachial artery and a large vein nearby. Then they sewed the brachial artery to the renal artery, and a vein in the arm to the renal vein. The clamps were released, and the kidney pinked up before their eyes. According to their reports, it also immediately began to produce urine. They tried to tuck it beneath a pocket in the patient's skin, but there was no space. Instead, they wrapped it in sterile sponges and covered it with rubber sheeting, leaving the tip of the ureter exposed.

Exactly how much urine the kidney made overnight is unclear. The day after the transplant, Hume and Hufnagel did have to cut back the ureter, as the tip was swelling and obstructing urine flow. This seemed to work. The young woman began to wake up, and two days after the transplant, she was fully alert.

At that point, Hume and his team noted, the transplanted kidney was petering out, and was removed. Remarkably, the patient's own kidneys began to recover—she was out of the woods. (Sadly, she died months later from an acute case of hepatitis she acquired from blood transfusions given prior to the kidney transplant.)

That same year, word was spreading around the world about Kolff's dialysis machines. George Thorn, the chief of medicine at Peter Brent Brigham Hospital, invited Kolff to visit. Although he didn't have any spare machines to give Thorn's team, Kolff did share the blueprints and drawings and presented his experiences with the devices to hospital faculty. Thorn assigned John Merrill, a young but aggressive medical doctor, to work on refining the plans for and building a dialysis machine to be used at "the Brigham," as the hospital was called. With some modifications, Merrill and his team constructed and tested their own version of Kolff's machine, and by 1950, they had performed thirty-three dialysis procedures in twenty-six patients. Hume was tasked with obtaining vascular access for these patients—for each run on the machine, he would have to identify an appropriate artery and vein for cannulation (i.e., the insertion of a needle into the vessel), which was no easy job. It wasn't until 1960 that more permanent techniques for dialysis were developed. Before this, doctors would search the patients' arms and legs to identify large veins and arteries they could cannulate with the big dialysis needles, sometimes cutting through the skin to get to them. After a few weeks the patients would run out of appropriate vessels, and dialysis would be discontinued. If their kidneys hadn't recovered, they would simply die.

Because of Hume's and Merrill's successes with hemodialysis, the number of young patients with renal failure seeking treatment increased dramatically. But other than a few sessions of dialysis, there was nothing to offer them. In 1948, Francis (Franny) D. Moore, at the age of thirty-five, became the chairman of surgery at the Brigham. Moore was already a famous surgeon, having conducted practice-changing research on burns and electrolyte management at the larger and more reputable Massachusetts General Hospital. He was a firm believer in applying science to clinical practice and was not afraid to offer risky surgeries or other treatments to improve patient care. Moore was also a key member of the maverick group being assembled to treat patients with end-stage renal disease.

In 1951 David Hume finished his training and was appointed by Moore, the head of the kidney transplant group at the Brigham, where he began his kidney transplant efforts in humans in earnest. For the medical community, David Hume was a surgeon in the mold of a James Dean, a sort of cultural icon who inspired a generation of transplant surgeons and whose personality remained electrifying despite horrible odds and public dissent about what he was trying to do. Hume couldn't accept that there were barriers to the things he wanted to accomplish. He was known to spend his daytime hours operating on patients and his nights working on animals in the lab, resembling a "human buzz saw."

Shortly after Hume's appointment as head of kidney transplant, Joseph Murray came on board and took over the efforts in animal experimentation, while Hume remained in charge of human transplantation. By this point, the Brigham had the key pieces in place for success. It had strong department heads in Thorn and Moore. It also had a functioning dialysis machine with an expert in dealing with the medical aspects of kidney failure in John Merrill. But the Brigham wasn't the only hospital that was getting into the game of kidney transplantation.

Most notably, two groups in France had embarked on their own transplant programs. One was run by the elegant and cultured urologist René Küss, and another by nephrologist Jean Hamburger. Getting kidneys for donation was no easy task. Brain death had not yet been defined. Nor was the practice of using kidneys from live donors a reality, given that there was virtually no reason to believe the transplant would work. Doctors would have to wait around for a patient to die in the hospital, and then quickly get consent and take the kidneys out, hoping they had a patient nearby to put them into.

For early transplants in France, these two groups decided to use kidneys from patients who had been sentenced to death and were to be decapitated by the guillotine. The donors gave consent, but this didn't make the surgeons or anyone else involved any more comfortable. In the words of Küss, "Despite the discomfort of the 'operation,' performed on the ground by torchlight, these kidneys were removed with great care and most of them were washed and perfused with Ringer's solution [i.e., a solution similar to saline] during transport in boxes especially designed for this purpose." Living donors were typically patients undergoing nephrectomy for some reason.

The French teams didn't have much success with these transplants. Most of the transplanted kidneys made urine for some time but ultimately stopped, and the patients succumbed to renal failure and death. Still, some important advances were made. First, Küss and the other French surgeons placed the kidney extraperitoneally (i.e., outside the bag in the abdomen), in the right side, sewing the renal artery and vein to the iliac vessels and plugging the ureter into the bladder, much as we do today. This was much easier than trying to place the kidney in the same spot where the old kidney sat, and much more practical than placing the kidney in the arm or the leg.

Another advance came from a transplant done by Hamburger's group. It was December 1952. A sixteen-year-old carpenter named Marius fell off a ladder, landing on his right side. He ended up in a small hospital outside Paris, where his injured right kidney was removed. Shortly after surgery, he was no longer making urine. He was transferred to the care of Hamburger at Necker Hospital in

Paris, where it was quickly confirmed that he had been born without a left kidney. He was as good as dead. Marius's mother pleaded with the doctors to take one of her kidneys and place it in her young son. This was a radical request at the time—never before had a kidney been taken from a healthy donor, who would get no medical benefit from the operation, to be used in a transplant that had virtually no chance of lasting long term. Hamburger's team did ensure that mother and son shared a compatible blood group—that much was known back then. As Hamburger wrote, "Our doubts and hesitation can be imagined, but finally we decided that to turn a deaf ear to the pleas of the family would be no less reprehensible than to agree." On Christmas Eve, the mother's left kidney was removed and placed in the right iliac fossa of Marius, similar to how Küss had performed the previous transplants. The kidney worked right away, and the team had high hopes. Marius improved clinically, and his labs normalized. But twenty-two days after transplant, the kidney stopped working. Wrote Hamburger, "This may have been, to start with, only a transplant crisis . . . that can be reversed with suitable treatment, but this knowledge lay in the future." True, the graft did work longer than expected, perhaps because of the genetic relationship between mother and child. Still, Marius died from renal failure shortly thereafter.

Boston, 1951–1954

David Hume's series of Boston transplants, which he wrote up in *The Journal of Clinical Investigation*, remains a fascinating and comprehensive summary of the experience of transplanting kidneys in humans without immunosuppression. Of the nine cases he was involved with in a three-year period, eight of the kidneys were placed in the right thigh of the recipient, as Hume thought this would be less traumatic for these sick patients, allowing monitoring of the kidney and ureter, and early identification and removal of a dying kidney. Hume knew these kidneys would not work long term; there was enough experience then in large-animal transplant and a smattering in humans that predicted this.

One major challenge was where to get donor kidneys. Given how new the practice of organ transplantation was, the general public was not familiar with the concept or even the possibility of donation. Hume had to find a source of healthy kidneys, and time their procurement so he would have recipients available. Then he learned of a procedure that had been developed by a surgeon on staff at the Brigham by the name of Donald Matson, whose specialty was the treatment of patients with hydrocephalus (or excessive accumulation of spinal fluid in the brain, causing swelling and pressure). Matson had developed a procedure whereby one end of a tube was placed into the ventricle in the brain and the other end threaded into the divided ureter, allowing the fluid to drain into the bladder. Matson, Hume learned, would remove and discard the kidney that used to be connected to the ureter. These discarded kidneys would serve as a perfect source for kidney transplants—until the Matson procedure was ultimately abandoned.

Five of the nine cases Hume performed were deemed failures. With the other four, however, he enjoyed some moderate success, and gleaned valuable information about kidney transplantation in man. First, he recognized that with deceased donors, the kidneys go through a period of anuria, failing to make urine for between eight and a half and nineteen days. The fact that Hume continued to follow these kidneys in the recipients, rather than remove them as failures, is critical and impressive. This period of anuria is something we still deal with today—we call it delayed graft function (DGF), or a "sleeper kidney," and about 30 percent of deceased-donor kidneys experience this, even with our improved preservation solution.

Hume documented the pathologic response of kidney rejection, and even identified a case of recurrent kidney disease in a transplant. He showed that the kidneys do seem to last longer in humans than had been the experience in animals. He showed the importance of blood group matching (as a couple of the kidneys he transplanted that didn't work were not matched). Although all his kidneys eventually failed and the patients died, there was one who stood out.

Case Number 9, as Hume called him, was a twenty-six-year-old South American doctor afflicted with Bright's disease, or chronic nephritis. A year before his transplant he had developed high blood pressure and leg swelling, and months before it his blood pressure worsened and was accompanied by headaches with visual changes. By the time he got to the Brigham, he had gross (i.e., visible) blood in his urine; was pale, anemic, and swollen; and was vomiting. A donor kidney became available from a young woman who had died on the operating table undergoing surgery for a narrow aortic valve.

As with the other eight transplants Hume performed, the left kidney was placed in the patient's thigh. But for this kidney, Hume came up with the idea of wrapping it in a sterile polyethylene bag and bringing the artery, vein, and ureter out through small holes in the bag. Why? Apparently, Hume thought that perhaps the factors in the serum and blood that had led to rejection of the other eight kidneys would not be able to permeate this bag. Once the kidney was perfused, the open end of the bag was sealed with heat from a cautery device. Skin grafts were needed to cover the wound despite the muscle and skin flaps Hume had constructed to do the job. The surgery itself went well, but the postoperative period was quite rocky. The patient bled extensively, receiving seven units of blood over eight days. His leg became massively swollen, and he was finally returned to the OR on that eighth day, in a "precarious state." His kidney transplant had made no urine, although his native kidneys did produce some. Massive amounts of blood were evacuated from his thigh, but the kidney looked healthy and perfused.

The kidney made no urine for eleven days, but then, on the twelfth day, it made about five milliliters. The urine output slowly picked up, and by the thirty-seventh day it was a whole liter. Thereafter, the kidney continued to produce between one and three liters of urine a day for six months, a normal, life-sustaining amount. On the eighty-first day after the transplant, the patient was discharged and went home feeling better than he had in years. He would come back to the hospital every two to three weeks for follow-up. Sadly, after almost six months, he became ill when he happened to be back at the Brigham for a follow-up visit. It is not entirely clear whether he died from infection, pulmonary emboli, or possibly chronic rejection of his graft. While Hume writes about this experience in the scientific paper without any real reflection on the emotions this setback might have prompted, Franny Moore gives a more explicit description of the effects it had on all those involved in the young doctor's care:

> Five months later he returned, his kidney now failing. He knew that he was going to die, but like so many patients who have had some but not complete success with surgery at the frontier of knowledge, he was grateful for the 6 months of life he had been given. The magnificent human spirit of such patients cannot fail to impress everybody

who sees them. He had a sort of calm assurance that the experience in his case would help others. Little did he (or we) know how right he was and how soon his prediction would be borne out . . . Our experience with this patient as much as any other single factor led to the successful initiation of kidney transplantation a little more than a year after his death.

Despite the ultimate failure of all the transplants in Hume's series, the one relative success left those involved with the belief that things would work out. In what would be a fateful episode that would forever change the field, Hume was called to active service in the navy, at the tail end of the Korean conflict. The timing couldn't have been worse for him. With his imminent departure, Franny Moore turned to Joseph Murray to take over the clinical transplant program.

The "Hump," Kurmitola, India, December 23, 1944

Charles Woods was preparing to fly the Hump, the name given to the route over the Himalayas that World War II pilots took to supply Chinese troops under the command of Chiang Kai-shek and American troops in China fighting the Japanese. He had made the flight numerous times before December 23, but this time he was training another pilot to see if he was ready to fly on his own. It turns out he wasn't, which Woods realized as the plane was accelerating along the runway in preparation for takeoff. Before the plane was able to leave the ground, the novice pilot abruptly put on the brakes. With the plane careening uncontrollably toward the end of the runway, Woods took over the stick and wrestled with it—to no avail. The plane fishtailed off the end and came to a stop after an encounter with a tree. Woods knew the plane was loaded with twenty-eight thousand pounds of fuel and figured it was going to explode.

I felt a first blast of heat, then my nerve endings must have seared because I lost all feeling. I knew what I had to do and I did it. I stayed calm and kept my eyes shut tight, hoping to protect them. I felt for the small Plexiglas window beside me, opened it and twisted through and slid down the fuselage. The plane was tipped over on its wing. I could hear the big old propeller still ticking over, and I knew I had to stay clear of that. I landed hands-first in a puddle of flaming gasoline, then ran until I could no longer sense the intense heat from the plane. Natives rushed out to help put out the fire that was consuming me. I discovered much later that they helped themselves to my watch and wallet as payment for their troubles.

Valley Forge General Hospital, Phoenixville, Pennsylvania,
Six Weeks Later

"When I first saw the young aviator, Charles Woods, he had no nose, eyelids, or ears, and his mouth—if you could call it that—was a raw opening." So begins the autobiography of Joe Murray. Murray was twenty-five years old, a recent graduate of Harvard Medical School (1943), and had just completed his surgical internship at the Brigham. He was now on active duty at Valley Forge, where he would spend three years as a staff surgeon before returning to the Brigham to finish his surgery training. Murray had known his entire life that he wanted to be a surgeon, but beyond that, he didn't know where his training would take him.

When Woods showed up in Pennsylvania, he was truly closer to death than life. He was dehydrated, malnourished, and infected, with 70 percent of his body burned and exposed to the elements. Knowing they had to find a way to cover Woods's burned areas, Murray and his team contacted the next of kin of a patient who had recently died and obtained consent to use his skin to try to save an injured pilot. Murray knew the skin would last only ten to fourteen days, at which point a new donor would need to be identified while they slowly tried to cover what they could using Woods's own skin. This would require multiple operations over many months, with many painful dressing changes for him to endure.

They embarked on Woods's care as a team, with each of them working on different parts of his body and face. "What we were doing was analogous to planting seeds, optimally preparing the 'soil' to accept and 'grow' the precious pieces of skin." Much to Murray's surprise, the deceased-donor skin stayed alive for a month or more. This was critical to Woods's outcome—it gave him time to improve medically and nutritionally and his donor sites time to heal so that new skin could be taken again. After eighteen months and twenty-four operations, Woods had finally healed adequately to go home. He had suffered incredibly but somehow maintained his sanity, positivity, and drive. He went on to be a successful businessman, politician, and great friend and supporter of Murray. Of Woods's case, Murray would write, "Charles was my introduction to the use of tissues from one person to save the life of another."

In 1951, Murray completed his training at the Brigham with some time in New York learning plastic surgery. He knew at that point that he wanted to be involved in surgical research, and he also saw the connection between his prolonged skin grafts on Charles Woods and the problem of making a kidney transplant work. He liked the seemingly insurmountable challenge, and the idea that these were dying patients who needed *something* to be done for them. Murray joined Hume's team, which was supported by Moore and Thorn. Dogs were the animal of choice. Most of Murray's work in those first three years involved removing both of a dog's kidneys and then transplanting one of them back into various parts of the same dog's body. His goal was to identify the best spot to sew in a kidney. Over time, Murray settled on the same spot that Küss had used in 1951 and Hamburger's team shortly thereafter.

Boston, Massachusetts, Fall 1954

"Like many twins, Richard and Ronald were best friends," said Murray of the Herricks, two identical twin brothers who would change the field of transplant. Richard and Ronald Herrick grew up in Rutland, Massachusetts, on a dairy farm. They both joined the military in 1950, at the outbreak of the Korean War, with Ronald serving in the army and Richard in the coast guard. At the time of their discharge in 1953, they were planning to move in together back in Massachusetts. Richard never showed up, and shortly thereafter Ronald received a letter that his twin was in Chicago being treated for chronic kidney disease. When Richard got sicker, he was transferred to Brighton, Massachusetts, where he could be closer to family when he died.

At the hospital, the twins' older brother, Van, asked the doctor if he might give his little brother a kidney. The doctor was about to say no when he realized he was looking at the spitting image of

Richard, in his identical twin, Ronald, standing there among Richard's family members. Maybe there was something that could be done. He knew that a crazy "bunch of fools" was doing research on this topic at the Brigham, and he reached out to John Merrill to see what he thought. At the very least, they could offer Richard Herrick some dialysis.

When Merrill heard about the twin, he was quick to accept this particular transfer. Given that it was Van, not Ronald, who had offered to donate a kidney, one might wonder what was going through Ronald's head at this point. "I had heard of such things," he said, "but it seemed in the realm of science fiction. For the first time, we began to feel the faintest glimmers of hope . . . I did some serious soul-searching. I mean, here I was, 23 years old, young and healthy, and they were going to cut me open and take out one of my organs. It was shocking even to consider the idea. I felt a real conflict of emotions. Of course I wanted to help my brother, but the only operation I'd ever had before was an appendectomy, and I hadn't much liked that." In the 1950s, there had never been a successful kidney transplant in humans (or animals), and all surgeries were considered high risk. But in the end, Ronald Herrick really had no choice.

Murray and his team knew that what they were considering doing was radical, that it went against the surgeon's oath to "first, do no harm." They obtained consultations from physicians, clergy, and legal experts, a majority of whom supported their moving forward—yet not everyone. Henry M. Fox, chief of psychiatry at the Brigham, was asked to see Ronald in consultation, and wrote this in his chart: "I think we have to be careful not to be too much swayed by our eagerness to carry out a kidney transplant successfully for the first time . . . The important question would seem to be whether we as physicians have the right to put the healthy twin under the pressure of being asked whether he is willing to make this sacrifice. I do not feel that we have this right in view of the potential danger to the healthy twin as well as the uncertainty of the outcome for this patient."

The situation was different from what we are faced with today. Patients with renal failure now have the option of long-term dialysis, so potential donors don't have to feel they are killing the patient if they decline to donate. Also, each potential recipient has access to more than one potential donor; with the advent of immunosuppression, donor and recipient no longer have to be identical. Most important, when we talk to potential donors today, we are able to discuss in great detail what to expect during the surgery, the recovery, and the long-term risks of donating a kidney. At the time that Ronald Herrick was making his decision, there really wasn't much data on how donating a kidney might affect one's life expectancy. He posed the question to Murray and the team, and was told, "We approached insurance companies for their actuarial tables and discovered that there was no increased risk from living with one kidney." In the end, it was truly a "leap of faith" for Ronald. In a lucid moment the night before the operation, Richard wrote him a note, "Get out of here and go home." At that point, Ronald was committed, and he tossed the note away without further consideration.

First, though, Murray and his team ran every test available to ensure that Richard and Ronald were indeed identical. This included sending them to the police station to get fingerprinted (which ended up blowing their cover: a newspaper reporter at the station caught wind of what was going on and published their story, raising the pressure on the surgical team). Murray also performed reciprocal skin grafts, and confirmed that after four weeks the grafts showed no signs of rejection and had fully taken.

Once everything was set to go, it occurred to Murray that he had never *done* a transplant in humans and needed to practice before operating on the twins. He contacted all the pathology departments in the city, and on December 20, a snowy night, he and Moore performed the first kidney transplant in the United States in humans in which the kidney was placed down in the iliac fossa (where we still place it today)—only both donor and recipient were dead, and one and the same.

Boston, December 23, 1954

J. Hartwell Harrison, the Brigham's chief of urology, performed Ronald's operation. Harrison had the most experience with nephrectomies (taking kidneys out) in the group, but he was also taking the biggest risk. He was the one performing surgery on someone who would get no (medical) benefit from the operation. As for Richard Herrick's surgery, Murray took that on his shoulders. His chairman, Moore, brought the kidney over from Harrison's OR after it was removed, and Murray proceeded to sew it in, connecting the artery and vein in one hour and five minutes (a bit longer than we take nowadays, but it was the first), with a total time of one hour and twenty-two minutes without blood flow. The kidney pinked up beautifully, and urine began to flow out of the ureter shortly thereafter. They had done it.

Richard Herrick went on to live eight more years, and in that time, he married and had children with one of the nurses who'd taken care of him. He ultimately died of renal failure, when his original kidney disease returned in the transplant.

As for the donor, Ronald Herrick, he died fifty-six years later, at the age of seventy-nine.

The response to this first successful transplant was massive. Stories about it were splashed all over the front pages of newspapers across the world, and discussions of kidney transplant filled the airwaves. Joseph Murray became an instant celebrity, and, amazingly, sets of identical twins with one in renal failure began flooding into the Brigham, including a thirteen-year-old pair and a seven-year-old pair.

In terms of lessons learned on the immune system, identical twin transplants didn't uncover anything useful or surprising, but they were a real shot in the arm for those few surgeons and scientists already trying to make transplantation a reality.

But where to next? Back to the animal labs. Peter Medawar had recently published his *Science* article demonstrating successful skin transplantation in genetically mismatched mice, a finding that electrified Murray and everyone else working in the field of transplant. It was time to find a way to make this work in humans.

Charlestown Labs, Massachusetts General Hospital, Boston

I was the new guy. I felt the sweat dripping down my back. I couldn't figure out why I felt so nervous. I had presented patients on rounds loads of times, and always enjoyed it. But something was different here.

After medical school, I had matched in a surgical residency at the University of Chicago, still thinking about that kidney and wondering what it would take for me to become a transplant surgeon. I had signed on for a five-year residency, but in my second year, I was so exhausted that I wondered if I could keep going. After learning that doing a few years of dedicated research during my residency could enhance my application to a transplant fellowship, I pretty much stumbled into this lab at the MGH (or "Man's Greatest Hospital," as they like to call it). The truth is, I had never done any research; I'd been a Russian language and literature major in college. Still, never one to let my own lack of skills stand in my way, I jumped at the opportunity to spend a few years in the premier transplant lab in the United States.

Looking around now, I felt massively unprepared. Present were five attendings, including the chief of cardiac transplant surgery, the chief of transplant surgery, the head of infectious diseases, the head of bone marrow transplant, and of course, the director, David Sachs.

"Okay, who's next?" David asked. "Josh, why don't you go?"

I walked over to the chalkboard and reached up to the magnet representing my first patient. "This is Zero-two-seven-eight-five," I said. "He received a combined heart and kidney transplant. His donor was radiated with one thousand rads. He was matched for class two, and received twelve days of tacrolimus. Kidney function is perfect, but his last biopsy showed some mild rejection."

I saw David motion me to stop. "Okay, back up here. Remind us again why the donor was irradiated? What is your hypothesis?"

I replied, "Okay, a little background. As you all know, kidney transplants across a class one mismatch, matched for class two, become tolerant after a short course of high-dose tac, but hearts do not. I am trying to prove that kidneys—"

"Stop," David said. "You don't prove, you *test*. You form a hypothesis and you *test* it with experiments that gather data with controls."

"Right, sorry. I am trying to *test* if kidneys have a population of cells that are radiosensitive, that traffic to the thymus and render the heart tolerant." There, got that out.

Then the head of bone marrow transplant spoke: "So, what can you tell me about the history of radiation in kidney transplantation in man? Have we used it before? Has it ever been used to achieve tolerance?"

I was pretty sure radiation had been used at some point, at least to treat rejection. "I know it was used," I said, "but I'll have to read about it more."

"Yes, good idea."

After all the patients had been presented, we walked down the hall and entered the ward to do bedside rounds. We started with those patients who had just undergone transplants: hearts, kidneys, and even a thymus and a spleen. We palpated their heartbeats, or looked at their urine output. It was just like the countless rounds I had been on over the last couple of years as a junior resident, with two differences. Multiple senior attendings had joined us, which was unusual, and the patients were of the four-legged variety: they were pigs, and their beds were cages. We could feel their heartbeats, since we had performed heterotopic heart transplants—we transplanted the hearts into their bellies as auxiliary hearts that we could monitor for rejection, without removing their own hearts. These can be palpated by simply putting your hand on the belly of the recipient.

Those three years in Boston opened my eyes to a world I never knew existed. In the pig lab we explored strategies to trick the recipient's immune system into accepting a transplant without the need for ongoing immunosuppression. That, in a nutshell, is tolerance, the very concept Sir Peter Medawar described in 1953. A tolerant recipient would have normal immune responses to other stimuli, just not to the transplanted graft. We were successful with a number of different strategies in the pig lab, including transplanting bone marrow cells along with the transplanted organ and transplanting the thymus of a donor along with another organ—allowing the donor thymus to reeducate the T cells in the recipient, preventing them from attacking the transplanted organ.

Franklin, Ohio, Nighttime, 1958

The story of nonidentical human transplant began in earnest in the middle of the night in rural Ohio when a surgeon removed an inflamed mess of tissue thinking it was an appendix. It was actually a kidney. Gladys was only thirty-one, a mother of two boys, married to John, a young roofer. Her husband had brought her to the emergency room when she developed pain in her belly. While examining her, the surgeon was impressed with her tenderness; she was clearly infected, and it looked for all the world like appendicitis.

Nowadays, Gladys would have been whisked off to the CT scanner and treated with antibiotics and IV fluids, and maybe everything would have been okay. But this was 1958, and CT scanners were still almost twenty years away. Without a CT scan, the surgeon wouldn't have known that Gladys had been born with only one kidney, and now she had none. In a few days she would be gone. But the surgeon had seen the news about the successful kidney transplant in Boston. Gladys's brothers convinced their boss at the Armco Steel Corporation to fly her there on the company plane, and Gladys made her way to the Peter Bent Brigham Hospital, under the care of Joe Murray.

In the years since Murray transplanted the Herrick twins, he was focusing his research on how to extend this phenomenon to the vast majority of patients with renal failure who did not have an identical twin. At the time, there were no immunosuppressive drugs available other than steroids, but there was one treatment that had been known to alter the immune system: radiation. A number of studies had used radiation successfully to transfer cells and cancers between animals, the same finding Alexis Carrel had considered just before the start of World War I but had never followed up on. Investigation into the potential relevance of radiation to the immune system was accelerated in 1945, after atomic bombs were dropped on Hiroshima and Nagasaki. John Merrill himself was a flight surgeon with the 509th Composite Group responsible for deployment of nuclear weapons and had studied the effects of radiation on survivors, many of whom had died from infection because their immune systems had been destroyed. He surmised that radiation might be able to play a role in kidney transplant.

Gladys was to be the first of twelve nonidentical kidney transplants to be performed with total-body irradiation to knock out the immune system and give the donor kidney time to "take," followed by a bone marrow infusion to restore immune function and prevent infection. The bone marrow was taken from various sources—the surgeons could not always obtain bone marrow from the kidney donor, so sometimes it was from a recipient family member. In the case of Gladys, shortly after her arrival, a kidney became available from a four-year-old patient undergoing the Matson procedure.

Gladys was kept in an operating room throughout her post-op recovery, as it was known she was at high risk for infection and this was the most sterile environment they could provide. After three weeks, the new kidney kicked in, and her dialysis sessions were discontinued. Gladys did well for a month, and both her care team and her family had high hopes. Ultimately, it was the bone marrow, and not the kidney, that failed. She became hopelessly infected and died.

Murray was crushed. He had stayed with Gladys day and night, and the loss devastated him. Yet his next case seemed even more hopeless: a young boy, age twelve. Like Gladys, he was irradiated before receiving donor bone marrow, but he died of infection before he could even get a kidney.

Then came case number three: John Riteris, who received a donor kidney on January 14, 1959. Because of the dramatic failures of the first two cases, with both patients dying from infection, Murray and his team altered their protocol to use sublethal, rather than lethal, irradiation. With this lower dose, and the close relationship between donor and recipient, fraternal twin brothers, no bone marrow was used. John Riteris was suffering from Bright's disease, and was close to death. After the transplant, the kidney kicked in right away, and initially things went well. Then, eleven days after the transplant, he fell ill from infection in his native kidneys. These were removed, and he slowly recovered. Over the ensuing months, John received booster doses of radiation and steroids, and ultimately, he recovered normal function. He lived with this kidney for twenty-nine years, with no immunosuppressive drugs and completely normal kidney function. His death was due to complications from heart surgery.

They had done it! A successful transplant across an immunologic barrier with normal long-term function. Still, despite this exhilarating success, the remaining patients in the trial of twelve all died fairly rapidly, of either rejection or infection.

This was a difficult time for all those involved, from the nurses to the residents to the surgeons. Although Murray sat with his patients day and night, and carried their heavy stories in his heart forever, he never considered stopping. "Some have wondered why we continued in the face of so many failed attempts," he would write in *Surgery of the Soul*. "With each failure, we learned a little bit more about preparing the patient, treating rejection, and timing the diagnostic tests. We were all focused on helping patients with end-stage renal disease. I was never discouraged. If we gave up, patients would have no hope at all . . . The identical twin experiences—by then 18 in number—were bright beacons leading the way toward our goal."

Enter Roy Calne

Oxford, 1958

"The lecture theatre was crammed full of students and graduates, a testimony to Medawar's celebrity in his field and his enormous power as a lecturer. The man stepped on to the platform, the buzz of chatter stilled, and he held the room spellbound with his brilliant oratory and extraordinary subject matter. Afterwards, a student asked if it was possible to apply the results of Medawar's research to the treatment of human patients. After a pause, Medawar said, 'Absolutely not.'"

This was just the kind of talk that would inspire one member of the audience, Roy Calne. He was always drawn to problems people told him he couldn't solve. When he was a medical student at Guy's Hospital in London, he took care of a young boy with renal failure. This was 1950, and Calne and his team had nothing to offer the boy, other than a bit of morphine and a bed to die in. This didn't sit right with Calne. Little did he know at that point, but others were already working to overturn this belief, especially David Hume and his team in Boston. Over the next eight years, Calne continued to train as a surgeon, always thinking about the possibility of transplant, and always frustrated with the negativity and learned helplessness regarding its potential, at least in England.

Then, in 1954, Joseph Murray performed his successful identical twin transplant. That and subsequent work going on in Boston had a major impact on both Calne and his teachers. Once he completed his training, Calne learned to perform kidney transplants in dogs and pigs at the Royal Free Hospital. Radiation treatment as a form of immunosuppression was all the rage, and he secured access to a cobalt irradiator, but quickly realized that radiation was not a realistic strategy for his patients. While searching the literature for some other way to suppress the immune system, he found an article in *Nature*, written by two hematologists, Robert Schwartz and William Dameshek, that reported the use of the drug 6-Mercaptopurine (6-MP) in rabbits to suppress an immune response to human serum. Calne decided to try it in dogs receiving kidney transplants, and it worked. He published this major finding in *The Lancet* in 1960, the first report of prolongation of kidney transplant survival in large animals using chemical immunosuppression. The paper gained Calne admission to the world of transplant immunology and an important relationship with Sir Peter Medawar, who insisted he go to Boston to work with the group at the Brigham. Medawar personally wrote a letter to Franny Moore, and shortly thereafter, Calne had a spot in the lab of Joe Murray.

When Calne and his young family disembarked from the *Queen Elizabeth* in New York City, he took a quick detour to the Burroughs

Wellcome laboratories in Tuckahoe, New York. There he met George Hitchings and Trudy Elion, two scientists (and future Nobel Prize winners) who had synthesized 6-MP along with other related chemicals that were developed for the treatment of cancer. These two scientists were impressed with Calne's work and informed him of some newer agents they thought might be even better. They gave him a few different derivatives of 6-MP to try, and he was on his way.

When Calne arrived at the laboratory of Joe Murray in 1960, he asked Murray if he could experiment with some of the drugs he'd gotten from Hitchings and Elion. Murray, who also was doubting the future of radiation, acquiesced. The two tested roughly twenty different drugs, and settled on B.W. 57-322, later to be called azathioprine, or Imuran, as the best agent. As Calne recalls, "The high point of these experiments was the presentation at the Brigham of the first canine patient to survive a kidney graft, treated with azathioprine, with normal kidney function at six months. After the case history had been read, the door was opened and my dog Lollipop pranced into the crowded auditorium, making friends with the distinguished professors in the front row." These victories in the lab, in addition to the intermittent injection of successes from identical twins, encouraged those involved in transplantation to continue moving forward. As Medawar remembered, "The whole period was a golden age of immunology, an age abounding in synthetic discoveries all over the world, a time we all thought it was good to be alive. We, who were working on these problems, all knew each other and met as often as we could to exchange ideas and hot news from the laboratory." After all these successes in the lab, the decision was made to move azathioprine into the clinic. Sadly, this was precisely when Calne was to return to England. He was disappointed to go but knew his path would continue in London.

After three consecutive failures with azathioprine, Murray transplanted Mel Doucette on April 5, 1962. A twenty-three-year-old accountant, Doucette was undergoing dialysis at the Brigham when a thirty-year-old man died on the operating table during heart surgery, and his kidneys became available. After the transplant, Doucette suffered a couple of rejection episodes, which were reversed with steroid pulses. Doucette's new kidney lasted twenty-one months, at which point he got a second transplant, which lasted an additional six months, until he died from hepatitis, which he likely contracted from his transplants or associated blood transfusions. This was years before there would be testing for the hepatitis virus. Given that this was 1962, that the kidney was from a deceased donor, and no radiation was used, this outcome was considered a massive success. According to Murray, "In just eight years, beginning with the day the Herrick twins walked through the doors of the Brigham in 1954, we had reached our goal . . . the successful transplantation of an organ from a dead donor—was now firmly in our grasp."

This was just the beginning for kidney transplantation. With a strategy finally in place to prevent rejection, chemical immunosuppression, surgeons now started the hard work of making kidney transplantation, from both living and deceased donors, a clinical reality. The number of centers devoted to transplantation began to grow, and those clinicians and scientists interested in it began meeting and vigorously presenting results and debating strategies. In addition to the Brigham, and Roy Calne in London, a few others joined the fray.

When he arrived at the University of Colorado in late 1961, Thomas Starzl, who is best known for his contributions to liver transplant, jumped wholeheartedly into the kidney transplant business. In his standard fashion, Starzl attacked kidney transplant both in the lab and clinically, using living and deceased donors. By 1963, he had presented and published his findings from more than thirty kidney transplants, obtaining better results than anyone had yet shown. His main contribution was in increasing the doses of steroids given in combination with azathioprine and also convincingly showing an ability to reverse rejection crises with high doses of steroid pulses.

By the mid-1960s there were more than twenty-five programs doing kidney transplants, and more and more long-term survivors. At the same time, there were still many failures. Late into the 1970s, the one-year survival rate for kidney transplant was no better than 50 percent. Literally half the patients transplanted would lose their grafts in the first year, many of them dying. If the 1960s was the decade of excitement in transplant, when it was finally shown that it could be done, the 1970s exposed the depressing reality that failure was as common as success.

The Penicillin of Transplantation

One drug changed everything. In 1958 the leaders of the Swiss drug company Sandoz (later, part of Novartis) set up a program by which employees traveling for business or pleasure collected soil samples to be screened for fungal metabolites that might have immunosuppressive or anticancer properties. Each week, twenty samples were tested. On January 31, 1972, sample 24-556, which had been collected by an employee vacationing in Norway, showed remarkable immunosuppressive abilities. Over the following year, the sample (named cyclosporine A, due to its cyclical structure and its derivation from fungal spores) was purified and studied. Jean-François

Borel and Hartmann Stähelin at Sandoz conducted extensive testing on it, and confirmed its impressive immunosuppressive properties.

Calne, who had spent the last twenty years searching for new compounds that could prevent graft rejection, attended a talk in 1977 with immunologist David White at the British Society for Immunology. There, Borel presented the findings on cyclosporine A. Calne and White were most impressed, and they managed to obtain a small amount of the drug to conduct some initial experiments in rat heart transplants. The results were too good to be true. After repeating the tests with similar results, Calne contacted Sandoz to obtain larger doses of cyclosporine A to test in large animals. He was told that Sandoz had abandoned its interest in the drug, but he could have what was left lying around the lab. With it, Calne performed kidney grafts in dogs and heterotopic heart grafts in pigs—again, with remarkable results. He was ready to move on to humans.

Calne flew to Basel, Switzerland, to convince the executives at Sandoz to commit to the production of cyclosporine. After much discussion, they "somewhat reluctantly agreed, believing that developing cyclosporine would be regarded as an ethical and humanitarian gesture but would probably be a money loser. They had no idea that this compound would revolutionize organ transplantation, create a huge new market, and become a gigantic source of revenue for their company."

Initial human trials, begun in 1978, were somewhat disappointing; high doses of the drug were toxic to the kidneys. Nevertheless, five of seven patients left the hospital with functioning grafts. Calne then performed transplants in thirty-four more patients using cyclosporine alone, and his results showed high patient mortality. Five patients had died, three had developed cancers, and almost none had normal kidney function.

Despite initial concerns about Calne's results, Starzl commenced, with his usual vigor, his own trial using cyclosporine. As he did with azathioprine before, he combined cyclosporine with steroids, which allowed him to decrease toxicity to the kidneys. Between December 1979 and September 1980, he treated sixty-six patients in a nonrandomized trial. Shortly thereafter, he moved to Pittsburgh and hit the ground running. In 1981, sixty-five more kidneys were transplanted using a prednisone-cyclosporine combination. The outcomes were unmistakably positive, with a one-year graft survival of 90 percent.

By 1983, cyclosporine had received FDA approval for kidney, liver, and heart transplant. The discovery of this drug was a huge step forward for the field of organ transplantation—perhaps on par with the major firsts that took place in the 1960s. Transplantation had arrived.

Part III

Expanding the Horizon Beyond the Simplicity of the Kidney

Only those who will risk going too far can possibly find out how far one can go.

—T. S. ELIOT

Open Heart

The Invention of Cardiopulmonary Bypass

This assemblage of metal, glass, electric motors, water baths, electrical switches, electromagnets, etc. looked for all the world like some ridiculous Rube Goldberg apparatus. Although the apparatus required infinite attention to detail it served us well and we were very proud of it. The heart-lung machines in use today bear as little resemblance to that early model as the jet plane of today bears to that magnificent conglomeration of wires, struts, and canvas that sailed into the air in 1905 from the dunes of Kitty Hawk with one of the Wright brothers at the controls.

—JOHN GIBBON JR., IN *A DREAM OF THE HEART*, BY HARRIS B. SHUMACKER

The path to kidney transplantation in humans followed a chronology from Carrel to Kolff to Medawar to Murray. Years before Murray and his colleagues unlocked the puzzle of kidney transplantation, entering the chest was anathema to most surgeons, and operating on a beating heart was a guaranteed massacre. The story of extrarenal transplants, starting with the heart, requires us to jump back in time, when a few brave souls thought they might be able to solve the mystery of the cardiovascular system.

My own story also returns to a time when my bravery was about to be tested for the first time.

University of Chicago, Internship and Residency

Internship was absolute chaos. I had arrived in Chicago after four years of medical school in New York with the official label of "doctor" but no actual skills, other than skinning dead people and cutting sutures in the OR. Aside from giving me a bunch of book knowledge about physiology and biochemistry, medical school had done very little to prepare me for what came next.

My first task as an intern was to gain experience taking care of sick people. The other interns and I covered so many patients and so many attendings that we could never get on top of the work. The more efficient I got with my tasks, the less I looked at patients as human beings and the more I looked at them as a list of boxes to check off. I inflicted so much pain on patients by sticking in needles and tubes, pulling out needles and tubes, cutting and sewing and draining, that I became desensitized to their experience. I would say, "Just relax, take a deep breath. You're gonna feel some pressure," but I wasn't really listening or paying attention. Patients would squirm around as I drained their abscesses, and I would just keep going. In fact, it got to the point where I regarded the patients as standing in the way of my accomplishing these tasks. I remember going into one patient's room to pull out a chest tube. He was sitting at a table with his food tray in front of him, but rather than get him back in the bed first, I pulled the tube out while he was sitting there. Box checked.

As uncaring as this sounds, some element of detachment is necessary in surgery. Once an operation starts, I completely disconnect from the fact that there is a person on the table, with a life and a family. Indeed, part of the reason a surgeon's training is so long and difficult is that you have to reach the point where you can do what needs to be done every day without becoming overly attached to, and therefore emotionally drained by, each case.

A couple of months into my training, I killed my first patient. My task was to place a central line in an elderly woman who needed IV antibiotics, a procedure I had done many times before. Placing a central line, also known as a central venous catheter, involves sticking a big needle into a vein in the neck or chest, and threading a catheter in fifteen to twenty centimeters so it sits in a vein right outside the heart, to administer medicines or fluids or withdraw blood. The patient was in her eighties, chronically on a ventilator, and with a tracheotomy in her neck. She didn't speak or interact in any appreciable way. And she was folded up like a pretzel. What was the point? I called the attending at home, and he said to put the line in. I gathered the kit and all the fixings and then called the patient's son. I explained that every procedure had risks, but this was relatively minor and shouldn't be a big deal. He thanked me.

I poured the betadine on the patient's chest, put on my sterile gloves, and took out the needle. I then put my thumb on her clavicle and plunged the needle into the skin below it. I hit the tip of the needle into the clavicle and then walked it down so I could advance it just below the clavicle and toward her chest, pulling back on the plunger the whole time. I advanced all the way in and didn't get any blood back. This meant I hadn't hit the vein. But I didn't get any air, either, so I figured I hadn't hit her lung, which was good. I pulled back, changed my angle, and advanced again. This time I got that sweet rush of dark red blood. I was in. I pulled off the plunger, and blood slowly poured back. I then advanced the wire and removed the needle. I threaded the triple lumen catheter over the wire and made sure to grasp the wire as it came out the back end, so it wouldn't get sucked into the patient due to the negative pressure going to the heart. I advanced the line about fifteen centimeters and removed the wire. All the ports drew blood.

I was just getting ready to sew the line in when the sat monitor, which keeps track of a patient's oxygen saturation, started to sound its alarm. I looked over at it: the level of oxygen in her blood was at only 40 percent. That's not good. It should be in the nineties at least. I quickly started sewing the line in, hoping that the monitor's alarm was going off because the sensor had slipped off her finger. Still, her saturation level kept dropping. Then her blood pressure cuff's alarm started sounding, and the sat monitor showed that her heart rate had dropped down into the twenties. She was about to code.

I realized that I must have nailed her lung on that first pass. As I finished tying the stitch in the line, I turned to the medical student who was observing this and said, "Open your book to 'tension pneumothorax' and tell me what gauge needle to poke in her chest and where to put it." At this point, I was listening to the patient's chest with a stethoscope; I couldn't hear any breath sounds. Not good.

The medical student yelled out, "Fourteen- or sixteen-gauge needle, second rib space."

I ran out of the room to the storage cart, yelling to a nurse, "Call a code!" From the cart, I grabbed a handful of big needles, ran back to the patient, and squirted some more betadine on her chest. Then I uncapped the biggest needle I had and plunged it into her chest, not far below her clavicle. A bunch of air rushed out. Good sign, I hoped. But the patient's sat monitor was picking up nothing.

I heard the overhead page in the background announcing the code. I knew within minutes that a bunch of people would pour into the room and ask me what I'd done. I plunged a second needle next to my first. More air. No other changes on the monitor. As I started doing chest compressions, people began coming into the room.

They stood there for a second, wondering what the hell I was doing. Then one of the medical residents went up to the head of the bed to start pumping breaths into the patient's trach tube with an Ambu bag.

I heard my chief resident, Charlie, behind me. "What happened?"

"I think I nailed her lung putting the line in."

"Shit. Let's just throw a big chest tube in and see if it works."

We both knew it wouldn't. He handed me a knife he always carried in his pocket—don't worry; it was sterile—and walked me through the tube insertion, since it was one of my first. We cut down over the rib in the fifth intercostal space. I poked through the muscle in between her ribs with a big clamp, and more air came out. We threaded the tube in the space between her rib cage and her lungs. During all this, the medicine team was performing CPR. We hooked the tube up to the suction canisters that the nurses had prepared. None of it made a difference.

About twenty minutes into the code, we all agreed to call it. She was dead. I had killed her. We call this a clean kill. No doubt who was responsible.

Charlie put his arm around me and said, "Don't worry about it. It's probably for the best."

One of the attending surgeons who happened to be around had come in the room. He came up to me and said, "She was probably dead already."

I knew she wasn't. I called her medical attending and let him know what had happened. He thanked me and apologized all at the same time. I offered to call the family, and he said that was a good idea.

I called the son who had given me his consent an hour before. "Hi, this is Dr. Mezrich, whom you talked to before about that line placement? I have some bad news about your mother. Unfortunately, there was a complication. Her lung was under a lot of pressure, and unfortunately the line nicked the lung—"

He cut in. "Did she die?"

"Yes. I'm so sorry."

He was quiet for a few seconds. Then he said, "Okay, thanks. We'll be in soon. Thanks for trying. She can rest now." He was trying to make it okay for me. I felt pissed off that I had been put in this situation but also incredibly guilty. I just had killed someone. That's not a feeling you get used to very quickly.

Toward the end of the second year of my residency, I worked for a particularly malignant chief resident—Charlie was his name (but not the same Charlie who helped me with the chest tube). At this point I was considered one of the best residents in the hospital, but what no one realized was that I had no idea how to do any operations. I was a likable guy, always comfortable telling jokes and interacting with the staff. Typically, the attending would conduct the operation, and I would assist, keeping everyone laughing. No one really focused on my lack of progress, and I figured at some point I would just start understanding how to do things. The reality is, it doesn't work that way.

One day, one of the staff told Charlie to take me through a case. It was something simple, like an inguinal hernia. Once the patient was prepped and draped and ready to go, Charlie just stood there and said, "Do it." But I hadn't read up on the procedure and had no idea what to do.

The next two hours were incredibly painful for me—at every step, Charlie demonstrated to me that I had no idea how to operate. And then, in an act of brutal honesty that actually changed the course of my training for the next decade, he said, "Wow, Mezrich, you really have no idea what you're doing in here. Everyone thinks you are this good resident, but I don't think you've learned anything about operating in the last two years. You are way behind everyone in your class."

I knew he was right, and deep down, his words made me nervous. So, I took them to heart. After that, no matter how late I got home or how early I had to get up the next morning, I prepared for the next day's cases. I bought a surgical atlas and went over the steps of each case. I tried to perform each operation in my mind. I was also more aggressive about trying to get in the OR, and I stopped joking around once there. Also, I honestly assessed how well I understood the cases after we were done. Surgical learning needs to be active. As much as I (and everybody else) feared Charlie, I owe him a big debt of gratitude. He woke me up to what it means to learn how to be a surgeon. If you are reading this, thanks, Charlie. Though you're still an asshole.

WHEN CONSIDERING ALL the services I rotated through as a resident, none was as crazy or as memorable as cardiac surgery. I often make fun of heart surgeons for the small area they operate on, and the simplicity of the heart as an organ, but the reality is I was very close to choosing cardiac surgery as a profession.

Cardiac surgery, to me, is very black and white. If you do a good job, the patients do fine. If you don't, they die. We used to say if the operation went well, you could stick the patients out in the woods and they would do fine. But if something went wrong, it didn't matter how hard you worked, you could never pull them through. This is an oversimplification, but it rings true in my mind. During my first cardiac rotation, as a second-year resident, the other resident who was supposed to be with me had just quit. So, I did the rotation alone, working upward of 130 hours a week.

I have so many crazy stories from those days and nights: opening chests in the middle of the night, placing huge lines and tubes, transfusing so many units of blood. When it comes to self-reliance, I think I built more of mine on that cardiac rotation than on any other service.

One particular surgeon stands out in my memory: Bob Karp. I learned more about physiology, patient care, accountability, and endurance from Karp than from anyone else in my training. A supremely well-trained heart surgeon, Karp got his medical degree back in 1958. He had a special interest in congenital heart surgery on babies and children, and he grew a large program in Chicago. Normally, a surgeon with the prestige of Karp would have fellows that would scrub in with him or her, but Dr. Karp did not have fellows, so the residents got to scrub in, even when he was doing crazy cases on babies, reconstructing their hearts, taking them from dysfunctional malformed bags and turning them into life-sustaining pumps. The operating room under Dr. Karp was kept silent, with the exception of barely audible classical music and his quiet commands to the technicians running the bypass pump. Karp was quite calm in the OR, but if you annoyed him he could absolutely destroy you.

Every night at 9:00 p.m. and every morning at 6:00 a.m. we residents were expected to make the "Karp call," during which we ran through our patient cases with him. You could never go home at night before making that call, and always had to be back in the next morning well in advance of the morning call. For twenty minutes before the call, we would scamper around the unit gathering our patients' big wall charts listing their vitals, labs, inputs and outputs, pressors, and meds. The nurses knew about the Karp call, and if they liked you, they'd have the charts ready, filled in, and maybe even stacked and good to go. If a chart was missing, the call was guaranteed to be a disaster. Each time I dialed Karp's number, my heart would be pounding.

"Hello," he would say quietly, almost as if he didn't know who was calling.

At first I tried a little small talk, but quickly realized this was not to be tolerated.

"Vitals! Output!" he would scream. He didn't want my editorializing.

At that point, I'd start hammering out the patients' numbers: blood pressure, heart rate, chest tube output, urine, pressors, labs.

At some point, he'd cut me off. "Give this, do that, transfer him out, give blood. Next." And I would move on.

Over time, Karp got to know me well. After about a month, he even let me give my opinion.

"Mr. Smith looks great, I'm not worried about him."

"Next."

Gaining Karp's trust was one of the proudest accomplishments of my life. By the end of my second month with him, I had decided

I wanted to go into cardiac surgery. Stopping and starting hearts, reconstructing misshapen pumps in little babies, the pure technical demand, the black-and-whiteness of it—it was all so intoxicating. Once Karp realized I was interested in the field, he would call me to his office for teaching sessions, where we'd review articles, talk about physiology, or he would ask me what my questions were. He knew everything about the heart, knew why some operations worked and some didn't, and he had no problem admitting when he'd made an error.

Though I didn't choose it in the end, I remain fascinated with the field of cardiac surgery, and much of that fascination is driven by the people who made open-heart surgery a reality. There is nothing cooler, more beautiful, really, than the perfectly orchestrated dance of going on bypass and shutting the heart off. When you open the chest and expose that pumping muscle that makes everything work, that bag of worms performing frenetic, cacophonic movements that are actually much more coordinated than you would first think; when you insert the various catheters that fill with blood and listen to the back-and-forth between the cardiac surgeon and the pump techs, working together perfectly in a give-and-take that seems so simple; when the cardiac surgeon says those fateful words, "On bypass," and runs the cardioplegia in, stopping the heart in its tracks; when you hear nothing other than the whir of the machine taking over the role of the heart and lungs—the mystery of the heart is now gone; it is just this bag of blood, so simple and easy to manipulate.

Of course, it wasn't always like that. And getting to this point, where surgeons start and stop the heart so many times a day and so easily, was perhaps one of the greatest achievements in medicine of the twentieth century. It was also a necessary step to even dream of one of the most extraordinary accomplishments in surgery: heart transplant. In much the same way that Kolff's dialysis machine was a prerequisite for success in kidney transplantation, the invention of cardiopulmonary bypass was an absolute requirement for heart transplantation, a feat made even more incredible by the personalities who dared to do it.

Operating Room, Massachusetts General Hospital,
Morning, October 1930

Every fifteen minutes, Jack Gibbon would inflate the blood pressure cuff on the patient's arm and listen closely to her pulse with his stethoscope as he slowly deflated the cuff. He had been doing this since 3:00 the afternoon before—seventeen straight hours of watching the heart rate, blood pressure, and respiratory rate of this young woman lying on an operating table, prepped and draped, with her arm poking out from under the sterile sheets. In the last few minutes, he had noticed that her breathing was more labored. Her pulse had become thready, and as he listened with his stethoscope, it was getting harder and harder to record a blood pressure. Finally, at around 8:00 a.m., the patient stopped breathing.

Dr. Churchill was notified, and he came bounding into the OR. "Within 6 minutes and 30 seconds Dr. Churchill opened the chest, incised the pulmonary artery, extracted a large pulmonary embolus, and closed the incised wound in the pulmonary artery with a lateral clamp." This surgery was called a Trendelenburg procedure, named after the surgeon who first performed it twenty-three years prior. No successful Trendelenburg procedure had been performed in the United States at that time, and that was not about to change on this particular morning.

All the efforts of Gibbon, Churchill, and the others involved proved futile, but an idea sprouted in Gibbon's head that years later would save more people than anyone could possibly imagine. "During that long night, watching helplessly the patient struggle for life as her blood became darker and her veins more distended, the idea naturally occurred to me that if it were possible to remove some of the blue blood from the patient's swollen veins, put oxygen into that blood and allow carbon dioxide to escape from it, and then inject continuously the now-red blood back into the patient's arteries, we might have saved her life. We would have bypassed the obstructing embolus and performed part of the work of the patient's heart and lungs outside the body."

As Gibbon thought about it more, he realized that it might actually be easier, as long as he was going to try to bypass the lungs and oxygenate the blood, to perform the work of the heart as well. He also realized that the tricky part would be not making a pump to replace the work of the heart but, rather, designing an oxygenator for the blood.

Gibbon got one other thing from his time in the lab at Mass General: a wife. Churchill's main lab tech, Maylie, spent the year helping Gibbon, and the two fell in love. The couple returned to Philadelphia for three years, where Gibbon set up his surgical practice and they started a family. For whatever reason, he remained obsessed with the idea of a heart-lung machine. He tried to stay involved in research, but with the time he was spending building a practice, he realized it would be impossible to get anything of substance done. He asked his old boss Churchill if he could return with Maylie and spend another year at MGH working on this project. Even though he thought it a fool's errand, Churchill agreed.

Gibbon returned to MGH in 1934 for a year of research. All his money for salary and supplies came from Harvard and MGH, and money was definitely tight. He bought an air pump in a secondhand shop for two dollars. He hand-made whatever he could from cheap parts he found lying around. He decided to focus on cats as his animal model, as he thought their small size and general availability made them the most reasonable choice. "I can recall prowling around Beacon Hill at night with some tuna fish as bait and a gunny sack to catch any of those stray alley cats which swarmed over Boston in those days. To indicate their number, the S.P.C.A. was killing 30,000 a year!" He got advice from various professors at the surrounding institutions, including MIT. The Gibbons came to the lab early in the morning and worked late into the night perfecting their device. Their plan was to take venous blood from a cannula threaded through an internal jugular vein into the superior vena cava of the heart of the cat. The superior vena cava is one of the big veins that return deoxygenated blood to the heart, where it would normally be pumped through the lungs to pick up oxygen and be sent back around the body. The Gibbons would then run this blood through their pump system, where carbon dioxide would diffuse off and oxygen would be pumped in. The blood would then be pumped back into a catheter threaded into the femoral artery in the leg of the cat and into the aorta, where it would circulate to the organs to deliver the oxygen needed for survival. Toward the end of the year at MGH, the Gibbons succeeded in performing life-sustaining cardiopulmonary bypass in a cat. As Gibbon recalls:

> I will never forget the day when we were able to screw down the clamp all the way, completely occluding the pulmonary artery, with the extracorporeal blood circuit in operation and with no change in the animal's blood pressure! My wife and I threw our arms around

each other and danced around the laboratory . . . Although it gives me great satisfaction to know that open-heart operations are being performed daily now all over the world, nothing in my life has duplicated the joy of that dance around the laboratory in the old Bullfinch Building in the Massachusetts General Hospital 32 years ago.

When the year came to an end, Gibbon returned to Philadelphia, more committed than ever to continuing his research while resuming a clinical practice. He made numerous improvements to his circuit, and by 1939 he reported long-term survival of cats after long runs of complete cardiopulmonary bypass. By 1941, he had improved his machine enough, in terms of size, reliability, and capacity, that he moved to the larger dog model and reported success again. With his goal of making bypass a reality in humans still in sight, Gibbon joined American efforts in World War II. He was a surgeon first, and came from a long line of prominent medical and military figures. As much as it pained him to put his project on hold, he spent the next four years in the military, much of it in the Pacific Theater. Thankfully, he lived through it, and his mind never strayed too far from the work that awaited him back home.

University of Chicago

I was staring down at the beating heart, feeling uncomfortable. When you are in the belly, sure, there can be small, pulsating blood vessels, and bowels moving around, gurgling at you gently, but the sheer violence of the beating heart seems to scream, "Get out!" I suppose it also bothered me that this patient was a seventeen-year-old. As a fourth-year surgical resident, I hadn't yet fully gotten to the point where I could separate myself from the patient during surgery. We were taking out the boy's left lung. He had previously had multiple wedge resections of his lung to remove a rare tumor that kept popping back up, and now the tumor was back with a vengeance. Everything had been going well up to this point. We had carefully dissected around his left pulmonary artery and had passed a big tie around it. Now we were ready to pass the jaws of a vascular stapler around it and fire. The stapler would lay down three lines of staples on each side and cut in between. It was a white load, so the staples were 2.5 millimeters in size. I knew that should be okay, but something was making me nervous. The pulmonary artery was such a big vessel, and the wall looked so thin—I could see the blood flowing through it with each beat of the patient's heart. Also the vessel was so distended. I slid the stapler around it.

"Fire it," the attending told me with confidence. I had operated with him dozens of times, and knew he was seasoned. I felt calmer.

I shut the stapler by squeezing the handle; it closed nicely. Then I flipped the button into firing mode and pumped the handle three times as the stapling mechanism advanced, cutting as it went. I flipped the stapler open and passed it back to the scrub tech. Everything looked fine, but then, slowly, a little blood appeared at the staple line, on the heart side. It was just a little oozing at first, almost imperceptible. I reached my hand down and gave it a small squeeze, just to test it. Suddenly it erupted. I had been leaning over, looking at it carefully, when it exploded with the force of a full beat of the heart. I slammed my hands almost blindly onto the heart, trying to stem the tide. I say "blindly" because my face was completely covered in warm, sticky blood. I could feel it all over my forehead, could taste it as it soaked my mask and dripped down my face, salty and metallic. I squeezed that bag of worms as hard as I could with both hands, reducing the eruption to a couple of small streams of blood with each heartbeat—kind of like when you put your thumb over the end of a garden hose to squirt your friends with a more focused stream of water. I looked up and saw, through the smears of blood coagulating on my glasses, the scrub tech, her hands next to mine, trying to subdue the blood escaping my grip. I noticed the blood trickling down her face and mask as well. I also noticed that my attending was no longer standing there. I swiveled my head and saw him lying on the floor behind me.

I looked back down at the field and wondered if there was anything I could possibly do. I lifted up one finger of my right hand to try to get a better look at how big the hole was, and was greeted by another spray of warm, sticky blood in my face. I clenched back down and yelled to the circulator, "Get the cardiac team! Overhead page them!" I stood there with the tech, her hands on top of mine, holding so tightly, for what seemed like an eternity.

Once the page went overhead, tons of people poured into this small operating room. When they saw the two of us with our faces covered in blood, they recoiled in horror. One of the cardiac surgeons and his fellow finally came in. The surgeon looked at us, looked down at the wound, and said, "Wow. Don't move." I could see a subtle smile come across his face.

My hands were shaking and sore, but I pushed the aching pain out of my mind. We still didn't have the bleeding perfectly controlled, but every time I tried to reposition my fingers, more blood squirted out. As we continued to grip the heart, the cardiac team scrubbed in next to us. They exposed the patient's groin, dumped betadine all over it, and slashed into his leg just as it came off his torso. They quickly dissected out his femoral vessels. I could hear lots of action behind me as the team wheeled in the bypass pump.

Once the cardiac surgeon had his cannulas in, the beautiful dia-
logue started between the techs and the heart surgeon.

Surgeon: "Ready to go on pump?"

Perfusionist: "Ready."

Surgeon: "Go on. Let me know when you're on full flow. Keep
maps at seventy mmHg . . . How's the drainage?"

Perfusionist: "Drainage good. Full flow."

And then, finally, mercilessly, "On bypass."

The blood filled the plastic tubes heading off to the bypass machine, turning them from clear to red in a matter of seconds. The
blood looked dark flowing out of the patient, and came back a
lighter red, filled with oxygen and devoid of carbon dioxide.

"Okay, you're good now," the cardiac surgeon said calmly. "We'll
take over." He added, chuckling, "You'd better go wash your face off."

I pulled my hands out of the chest. Although the heart was still
beating, there was almost no blood flowing out now. It was all going
into the machine. At that moment, I thought, *I am definitely going
into cardiac surgery. These guys are gods.* If you can shut the heart off
and turn it back on at will, what *can't* you do?

JACK GIBBON, UPON returning from the war, had lost none of
his obsession with his bypass machine. If anything, he was more
motivated than ever. Despite an inability to stop the heart, courageous surgeons had begun serious efforts at closed heart surgery to
address the cardiac trauma they encountered during the war; this
essentially meant blindly shoving a finger into a hole in the heart
caused by shrapnel and trying to sew it up before the torrents of
blood pouring out led to the death of the patient. Once the war
ended, these same courageous surgeons tried to blindly take on

abnormal heart valves destroyed by rheumatic fever. While there were a few successes, mortality was absurdly high.

Gibbon knew there had to be a better way, but he needed a lucky break to get over the hump of designing something efficient enough to support a human. He had recently been appointed professor of surgery and director of surgical research at the Jefferson Medical College in Philadelphia, where a first-year medical student happened to become interested in his efforts. It just so happened that this medical student's fiancée's father was close friends with Thomas Watson, chairman of the board of directors of IBM. A meeting was arranged, which Gibbon later described:

> *I shall never forget the first time I met Mr. Watson at his office in New York City. He came into the anteroom where I sat, carrying reprints of my publications. He shook his head and sat down beside me. He said that the idea was interesting and asked what he could do to help. I remember replying rather bluntly that I did not want him to make any money from the idea, nor did I wish to make any money from it. He said, "Don't worry about that." I then explained that what I needed was engineering help in the design and construction of a heart-lung machine large and efficient enough to be used on human patients. "Certainly. You name the time and place and I shall have engineers there to discuss the matter with you." From that time on we not only had the engineering help always available, but IBM paid the entire cost of construction of the various machines with which we carried on the work for the next seven years.*

Both Gibbon and IBM held true to their promise not to make money on this endeavor.

Working together, Gibbon and his newfound engineers made

dramatic improvements to their device, particularly on the efficiency of blood flow and the ability to oxygenate it. The engineers found that creating turbulence in the flow of blood could dramatically increase the blood's oxygenation, which was good because, prior to that, Gibbon had thought the pump would have to be seven stories tall. They designed large screens through which the blood would flow, causing both turbulence and increased surface area through which the oxygen could diffuse into the blood cells. With every advance made in the design, extensive tests were performed in dogs. At first there were many deaths, but each death led to an improvement. By 1952, Gibbon felt ready to test his device in humans.

University of Minnesota

It is virtually impossible to talk about the origins of open-heart surgery or heart transplantation without talking about Minnesota. There wasn't much of a surgery program there until the arrival of Owen H. Wangensteen, who became chair in 1930 and remained so until his retirement in 1967. Wangensteen believed in big surgeries and thought surgeons needed to be scientists, too. He required all residents to conduct research and obtain PhDs during their training, unusual in those days. He also had a requirement that residents master at least two foreign languages before they graduated. In his thirty-seven years as chairman, Wangensteen took the University of Minnesota Department of Surgery from a small rural program to one of the finest academic departments in the country.

Wangensteen had an eye for recruiting the best people and a talent for helping them pick projects and find funding to carry out experiments. And for whatever reason, he amassed the most impressive group of cardiac surgeons the field might ever see at one institution. This was probably because the heart was the new frontier in surgery at that time, and Wangensteen was one of the few people who would support these crazy guys. The stories of the heart surgeons at Minnesota in the 1950s could fill multiple books (and have), but one surgeon in particular deserves mention, C. Walton Lillehei. Of all the pioneers in surgery, Lillehei has to be the most fascinating, daring, inspiring, and complex character. He did his surgical training at Minnesota, where he became a favorite of Wangensteen's. (He also had the honor of being fileted by Wangensteen after he was diagnosed with a lymphosarcoma right at the end of his residency—the surgery involved a neck dissection and a sternotomy [opening of the chest] with removal of multiple lymph nodes. Between Wangensteen's bloody attack on the cancer and multiple sessions of radiotherapy, Lillehei was cured.)

Lillehei thought the bypass machine Gibbon was developing was just too complex to be a realistic option. Too many people were involved, and too many moving parts exposed the teams to myriad mechanical errors. He had a different idea, one he first tried out in the animal lab with the help of some of his residents. Why not use another animal to serve as a bypass machine? They could hook up another dog, using a catheter in an artery and a vein, with tubing connecting these catheters to a vein to draw blood from the subject (i.e., the dog whose heart is being bypassed) into the "pump" (the second dog) and another set of tubing to return the oxygenated blood from the artery of the "pump" into the subject. Lillehei and his team spent the year experimenting on sets of dogs, defining appropriate flows, creating and fixing ever-more-complex defects that they themselves had created in the hearts of these animals. Once they got the details worked out, they found they could easily have the heart open for more than thirty minutes on cross-circulation, and still the animal would wake up and act normal.

By the beginning of 1954, Lillehei was ready to try this out on a human—or two humans, really. He didn't want to fix a small defect, such as an atrial septal defect (ASD). Too many people were having success doing that by then, using severe hypothermia (packing people in ice), which allowed surgeons to slow the heart for short periods of a few minutes. Wanting to tackle something more challenging, something that no one else had been successful with, he began looking for a child with a *ventricular* septal defect, or VSD. The ventricles are much thicker chambers than the atria, more complex and challenging.

It did worry Lillehei that this would be one of the first examples of an operation during which someone who didn't need surgery would be put at risk—this was shortly before the first living-donor kidney transplant in the United States was attempted—and for which the risk to the donor was unclear. The donor would simply have a small incision on his upper leg, with a catheter placed in his femoral artery and femoral vein. After the procedure, these vessels would be repaired and the wound closed. But heparin would need to be given to prevent clotting of the circuit. At any point in the procedure, the lines could become dislodged, leading to bleeding. Also, the donor would be exposed to whatever infection the "recipient" might have—but these were all babies or young children, so Lillehei hoped that risk was low. Perhaps the biggest risk would be the potential for air to get into the circuit. He and his team would need to be vigilant.

In March 1954, Lillehei identified his first patient, Gregory Glidden, a one-year-old boy with a loud heart murmur and an enlarged heart. By the time Lillehei looked at Gregory's angiogram, it was clear that the child had a VSD and didn't have much time left. Lillehei told Gregory's parents, Lyman and Frances, about his plans. He would need to check their blood types, and if they were a match, he would hook one of them up to Gregory in the OR. He was clear that there were no guarantees, and that he had never done this before on a human being. The parents agreed immediately; they would do anything to save their son.

After Gregory was asleep, they opened his chest and looked at his heart. Everything was a go. They placed a cannula in his aorta and one in his superior vena cava. The boy's father, Lyman, was wheeled into the room, groggy but awake. He looked over at his son, and Lillehei told him everything looked good. They put Lyman to sleep and then cut down into his thigh to expose his vessels. Once everything was hooked up, and all the lines were examined, Lillehei gave the signal: "Pump on." The small roller pump whirred to life, and blood flowed through the lines. There was no leaking. Lillehei placed tourniquets on the inferior and superior vena cava and pulmonary artery and cinched them down. Gregory's heart continued to beat, but it was not getting any blood. Lillehei cut into the ventricle. When he got inside, there was still bleeding, more than he expected, but they could keep up with it with suction. Lillehei carefully and accurately sewed up the defect with interrupted sutures. He then filled the heart with saline to get rid of the air and closed the wall. As he released the tourniquets, he asked for the time: fifteen minutes and twenty seconds. They had done it!

Sadly, the intraoperative success was not matched with a good outcome. Gregory did great at first, but then became infected with postoperative pneumonia. Despite heroic efforts from Lillehei and his team, they lost him on post-op day eleven. Lillehei convinced the child's parents to let him do an autopsy, during which he was at least happy to see that his repair was intact.

Lillehei pushed on with cross-circulation. He performed a total of forty-five operations; twenty-eight of these children survived. The rest died, a reality that hung heavily on him. Still, he persisted. Whenever he did have a death, he always told the family himself—not something all cardiac surgeons did in those days. While at first he limited the surgery to VSDs, as he got more comfortable, he became the first person to correct more complex congenital anomalies, including a tetralogy of Fallot and an atrioventricular canal, heart defects whose repair required significant effort and a lot of imagination, given that no one else in the world had ever repaired them.

Maybe the most dramatic case of the series was the first tetralogy Lillehei fixed. The boy, a ten-year-old, had blood type AB, which no one in his family shared. The Red Cross went through its records and identified Howard Holtz, a twenty-nine-year-old highway worker and father of three. They approached him to see if he would be willing to save a young boy he'd never met. Holtz agreed. "I just wanted to do what I hope someone would do if my child were a blue baby." The operation was a success.

Also among Lillehei's cases was one disaster. A young woman, Geraldine Thompson, was to serve as the "pump" for her eight-year-old daughter, Leslie, who was dying from a VSD. After Leslie's chest was opened and the surgeons were getting ready to operate on Geraldine, an IV running into Leslie for her anesthesia inadvertently got filled with air and pumped a major air embolus into Geraldine's brain. The mistake was recognized and the surgery aborted. Lillehei closed Leslie's chest without fixing her defect, and she died a few years later from her VSD. As for Geraldine, she lived the remainder of her life severely disabled, requiring a long-term care facility (although she lived to the age of eighty-eight). Even though Lillehei himself hadn't made the error, he felt responsible, and he urged the family to sue him to help with the costs of Geraldine's care. (They did—and lost the suit.)

Lillehei presented his results at a few major conferences, with mixed responses. At this point the only person in the world performing open-heart surgery, he was treated like a hero by some in the field, but others found him entirely unethical. True, he didn't know what the exact risk was for his "donors"; then again, neither had Murray or Hume or others when they'd started doing transplants. There were no other options for their patients, and with his operation, Lillehei gave them a better-than-50-percent chance of survival. Who among us wouldn't have jumped at the chance to save our baby or young child if given this option?

Back to Philly, 1952

"I believe we are approaching the time when extracorporeal blood circuits of the heart-lung type can be safely employed in the treatment of human patients." This is what Jack Gibbon said at the annual meeting of the American Association for Thoracic Surgery. By this point, Gibbon and his team had achieved great success with their optimized Gibbon-IBM machine. They had performed numerous successful pump runs on dogs, where the team put the dogs on bypass and opened either the right atrium or the even more challenging right ventricle. They had explored the valves, and had caused and repaired injuries throughout the heart. They had per-

formed multiple runs up to thirty minutes, thought to be of sufficient time to fix the majority of defects they were aware of at that
time, and the dogs had woken up and been fine.

Gibbon's first attempt at full cardiopulmonary bypass in a human was in February 1952 at the Pennsylvania Hospital. The patient was a fifteen-month-old girl, extremely ill, weighing eleven
pounds, most of which was water weight from fluid overload. The
preoperative diagnosis was a large ASD based on examination, but
cardiac catheterization was unsuccessful, likely due to the child's
size. Gibbon took her to the OR and placed her on his machine. The
heart was severely enlarged and beating poorly. When he opened
her right atrium, he found no ASD. He closed the heart back up, but
it couldn't take over and liberate her from the machine. She died on
the table. The autopsy showed in fact a different diagnosis: a patent
ductus arteriosus, a congenital abnormality *outside* the heart, which
could have been fixed in those days without a bypass machine. Who
knows if the girl would have survived if this had been known preoperatively.

The following year, in May 1953, Gibbon treated Cecilia Bavolek,
an eighteen-year-old in her first year of college. When she was a
baby, her parents were told she might have a heart defect, based
on the presence of a murmur, but she thrived and needed no treatment. She started to become ill at the age of fifteen, and over the
next three years she was in and out of the hospital with shortness
of breath, fatigue, fluid overload, and related symptoms. She was
ultimately diagnosed with a likely ASD, although VSD remained a
possibility. Gibbon agreed with the diagnosis, and Cecilia was taken
to the OR and placed on full bypass. Gibbon opened her swollen
heart through her right atrium, and indeed encountered an ASD.
He explored her ventricle and found no defect. He carefully sewed

up her atrium, and then closed her heart. The machine had served as Cecilia's heart and lungs for twenty-six minutes.

That July, Gibbon attempted two more cases, both five-year-old girls. The first one was correctly diagnosed with an ASD, but her heart was so weak that after repairing it, Gibbon could not get her off the bypass pump. After trying for four hours, he gave up, and she died. When he opened the heart of the next little girl, he indeed found an ASD, but there was also a VSD and a patent ductus arteriosus. He attempted to close the defects, but the bleeding was too severe, and she bled out on the table.

That was it for Gibbon; he was done. He declared a moratorium on his use of the machine and never returned to heart surgery. This was shocking to the heart surgeons of that era. How could someone who had spent more than twenty years working on this machine, who was about to summit his Everest, just walk away at the relatively young age of forty-nine? And given that he did, without doing the heavy lifting of moving bypass from a conceptual success to a clinical reality, does he deserve credit as the father of cardiopulmonary bypass, much less of open-heart surgery?

Gibbon continued as chairman of surgery at Jefferson, doing thoracic and abdominal surgery until retiring at the age of sixty-three. He had many other interests, including poetry, painting, travel, and of course his beloved family. He left the development of the cardiopulmonary bypass pump and open-heart surgery to others—primarily Lillehei, and John Kirklin at the Mayo Clinic. Kirklin had visited Gibbon in 1952 and was interested in making his own version of the pump. Gibbon had the blueprints sent to Mayo, and with the help of engineers from IBM and Mayo, updates were included in a version of the machine made there. Kirklin pushed forward, reporting his own somewhat successful trial of eight patients in 1955. Four of the eight survived.

By this point, Lillehei could see that cross-circulation would ultimately be abandoned once bypass became more predictable. He hated the complexity of the machine, and the cost. Then, in 1954, Richard DeWall, who had joined his team, set out to design a simpler, cheaper bypass machine. His strategy for oxygenating the blood was to create a reservoir for blood where oxygen was bubbled in at the bottom. It was simple and efficient. Lillehei used this machine successfully for the first time in May 1955.

Also that year, the great Willem Kolff got into the bypass game. When Kolff was working on his kidney dialysis machine, he had recognized that the blood seemed to become a lighter red as it ran through his sausage casings. It occurred to him that perhaps if he surrounded his membranes with high-pressure oxygen, this would diffuse across the membrane and oxygenate the blood. Once he was at Cleveland, he worked with others there to perfect this device, to be called a membrane oxygenator.

Over the next decade, more and more heart surgeons obtained machines using these three techniques, and got better and better at operating them. A version of the Minnesota machine was marketed for as little as a hundred dollars, compared to thousands for the more complex Gibbon version. Although each version had success, over time the membrane oxygenator won out. Regardless, these three machines and the efforts of these men made open-heart surgery a reality. By the end of the 1950s, open-heart surgery was commonplace, and increasingly complex surgeries were being attempted. How much further could the field go?

| 7 |

Hearts on Fire

Making Heart Transplant a Reality

Our greatest glory is not in never falling, but in rising every
time we fall.

—CONFUCIUS

For a dying man, it is not a difficult decision [to accept a heart
transplant] because he knows he is at the end. If a lion chases
you to the bank of a river filled with crocodiles, you will leap into
the water convinced you have a chance to swim to the other side.
But you would never accept such odds if there were no lion.

—CHRISTIAAN BARNARD, *ONE LIFE*

October 2006

It started with some pain and numbness in her right arm. It didn't
seem like a big deal at first; it probably was related to her busy
day: the long walk, the house cleaning, and of course the never-

ending tasks that come with a new baby. Tina had just had her son (her third child) two months before, and maybe she had pushed things a bit. Still, something just didn't seem right, and although she couldn't quite put her finger on it, she kept trying to get an image out of her mind: her three children without their mother.

When she woke up in the middle of the night, her illness was undeniable. She still had the arm pain, and now she was experiencing nausea and diarrhea. Her husband called an ambulance—he wasn't going to take any chances—and after a little time in the hospital, Tina felt much better. The nurse told her that as soon as her last blood test came back, she could go home. That sounded good. She almost felt embarrassed that she'd come to the hospital in the first place, for a little arm pain and an upset stomach. Then the nurse came back to the room and told her there was some news. Her blood test, that last one they'd done, had come back abnormal. It indicated that Tina had suffered a heart attack. What? She hadn't had any chest pain, had always been healthy, and had never smoked. There must have been a mistake. They told her they needed to transfer her to a bigger hospital, in Lacrosse. After making some calls to her family, she was packed up into another ambulance and whisked out the door.

After she was settled in a room in Lacrosse, she told the nurse there she had to go to the bathroom. She distinctly remembers getting up, again feeling like something was wrong—and then she collapsed. She was in the throes of another major heart attack. Things got a bit hazy after that. She was told she was brought back to life twice. She remembers waking up to see her family gathered around her bed looking worried. She remembers being flown in a helicopter to Madison, and two nurses holding her hand.

Things progressed very quickly for Tina. She was diagnosed with peripartum cardiomyopathy, or PPCM, a rare disease in which, during pregnancy, a woman's heart gets stretched and becomes dilated. The muscle weakens, which leads to symptoms of heart failure, including leg swelling, shortness of breath, fatigue, stroke, and even heart attack. The surgeons placed an LVAD, or left ventricular assist device (that is, a mechanical pump implanted into the left ventricle), in her heart. The device was hooked to wires that came out of her belly and connected to battery packs, which Tina ultimately would carry around in a backpack.

Tina spent about a month in the hospital, but with the help of the LVAD, she was able to recover. Before she was discharged, doctors taught her how to take care of the LVAD: how to change the batteries, how to hand-pump the device if the batteries ran out. She was sure that would never happen to her.

But it did, and she ended up back at Madison. Somehow Tina hung in there, with a failing heart and a new baby. She had been placed on the heart transplant list and needed to get a heart before another event killed her or the LVAD became infected. It was a race against the clock, but she made the best of it, despite sleeping most of the day and night.

"Even at a year and a half, my son knew that no matter when we left the house, he'd grab that pack of all the batteries," she said. Tina was on the waiting list for fourteen months. Toward the end of that time, she got an infection of her LVAD, which was life-threatening but also moved her up the list.

On the morning of December 5, Tina's phone rang. She was getting her kids ready for school. It was seven o'clock in the morning, with a heavy snowstorm outside. She wasn't expecting a call that early and was surprised by a voice she didn't know. It was one of the coordinators at the cardiac program at the University of Wisconsin: They had a heart for her. Was she ready?

She felt the mix of exhilaration and guilt that all our patients experience. They wait so long for that call, hoping against hope that it will come before it's too late. At the same time, they know they are waiting for someone else's death, someone they will never meet but to whom they will be connected in a more intimate way than their parents, children, or lovers—and for the rest of their lives. Tina remembers looking outside and noting that the blizzard would make the long drive to Madison treacherous.

As it turned out, this same blizzard, the biggest of the year, is what had killed Katie. She was just twenty-four and had four children under the age of six. Driving a car, Katie took a curve at just the wrong time. A snowplow was coming the other way. A car driving behind the oncoming plow slid out of its lane, colliding with Katie's car. The only good to come out of this tragedy was that Katie had actually just talked to her mother a couple of nights before about her wish to be an organ donor if she died. She hadn't had time to sign up for it yet, and it had certainly never occurred to her that doing so was an urgent matter. That desire would turn out to be a gift to her mother and to Tina.

Tina had been so strong throughout the fourteen months on the waiting list, always staying positive, but she finally broke down when she was being wheeled to the holding area of the operating room right before the surgery. She started to think about dying, about her family—and about the donor. Her surgeon, Dr. Takushi Kohmoto, held her hand and, in his calm and quiet way, told her everything would be okay.

Her surgery was long, complicated by the LVAD that had been in there for over a year. Still, everything went well, and Tina's recovery was as smooth as could be. She was out of the hospital in less than a week. And since that time, she has been entirely healthy. She has been a mother to her three children, including her little boy, who is not that little anymore. She calls him her "little miracle" because the cardiomyopathy that resulted from the pregnancy connected her to the miracle of heart transplant. And what a miracle it is. Patients come in and get this simple operation, and it turns their life from a living hell to a normal existence overnight.

"I cherish every day. I mean, I've always been a positive, upbeat person," she says, adding that she is definitely more so now. "I'm thankful," she says. "I don't get upset over little things . . . I just cherish everything I have." And with her oldest child getting married in October, she says, "I get to be here to see that." Tina has seen pictures of Katie, her heart's previous owner, and has gotten to know Katie's mother. She is Facebook friends with Katie's children, and hopes someday to meet them. She knows the relationship with Katie's family has been healing for her and hopes the same is true for them. She celebrates Katie's birthday every year, almost as if it were her own.

At the end of our conversation, I asked Tina if she knew about the incredible history of heart transplantation. Not really, she told me, but she would love to learn about it.

Back to Minnesota, 1949

Norman Shumway arrived for his surgical internship at the University of Minnesota almost by chance. He'd grown up in Michigan and entered the University of Michigan with plans to become a lawyer. But this was 1943, and the world had different ideas. After only a year of college, he found his way into the military, and given how well he did on his IQ test, he was sent to engineering school. After six months, though, the military decided it was short on doctors and gave Shumway a medical aptitude test. He performed so well that he soon found himself at Vanderbilt Medical School.

At Minnesota, Shumway spent time in the lab experimenting with hypothermia on dogs. He was fascinated by the power of the cold to slow the heartbeat and decrease the physiologic demands of the body and the brain. After two years of training, Shumway spent two years in the air force during the Korean War. Upon his return to Minnesota, Lillehei was working in the lab with his cross-circulation experiments, and when he was ready to use this technique to operate on Gregory Glidden, Shumway was the resident assistant. He was also present at the introduction of the DeWall-Lillehei bubble oxygenator, and took part in numerous open-heart cases using bypass. By the time he finished his training in 1957, he was ready to start a cardiac surgery program in the real world. He had thoughts about how he could combine hypothermia with cardiopulmonary bypass to make the whole thing safer. That was his plan.

While Shumway was still a resident, another young trainee showed up in Minnesota. He was not from the Midwest, not even from North America. He had never spent a winter in the frozen tundra that Shumway took for granted. His name was Christiaan Barnard, and he was from Cape Town. Barnard had heard of the legendary program sprouting up at Minnesota and knew that no training like this existed in his native South Africa.

When Barnard arrived in December 1955, Wangensteen put him to work in his laboratory studying the esophagus, a challenging organ to operate on. In the lab next door, Lillehei's resident was working on a project with his cardiopulmonary bypass machine. This was the first time Barnard had ever seen the machine, and he was immediately fascinated. He started helping out with that project, and before long, the resident asked if Barnard would like to assist the next time they used the pump on a human. Years later, Barnard described that day: "[E]ven now I can recall the details of that morning, the first time I witnessed the life of a human being held in a coil of plastic tubes and a whirling pump." His description of that first open-heart surgery sounds like a religious experience.

The bypass machine was switched on, and Barnard watched the dark blood flow into the chamber of bubbling oxygen and the light-colored blood flow back out. "[T]he longer it ran," he would later write, "the more exciting it became. This was more than a machine. It was the gateway to surgery beyond anything yet known. While it stood in for heart and lungs, vast repairs could be made inside the body. New valves could be put into the heart, maybe even a whole heart itself."

In this moment, Barnard knew what he wanted to do with his life. He met with Wangensteen the next day, to tell him of his interest in completing a full training program at Minnesota with a focus on surgery of the heart. Wangensteen was happy with that and told Barnard it would take six years. But with a family back in Cape Town and not much money, Barnard was eager to move on. He told Wangensteen he could do it in two. Wangensteen thought this impossible—two years on the clinical service was required, and in addition, he needed to do experiments, write a thesis, and be fluent in two languages besides English. Barnard promised him he could do it. He had already done loads of research that could serve as the basis for some future work. He knew Afrikaans, so he felt he could pick up Dutch and German quickly. His plan was to work on the wards during the day, work in the lab at night, and squeeze in the languages and the writing of his thesis on the side.

"When will you sleep?" Wangensteen asked.

"I don't need much," Barnard told him.

"All right. Let's see what happens."

Over the next two years, Barnard worked like an absolute dog. He worked in the clinics, worked in the lab, and still found time to sleep with a bunch of nurses while his family was back in Cape Town. To save time, when Barnard would return to his apartment well after midnight, he would walk into the shower with his clothes on and then hang them to dry so he could put them back on in the morning. His training included an eleven-month stint on Lillehei's service, probably the most important experience of his career. Unlike Wangensteen, who was a great chairman but rarely let his assistants do anything in the OR other than hold retractors, Lillehei entrusted his residents to perform independently. This came with great responsibility.

Barnard describes one day early in his training when he was opening a young boy's chest and preparing him for bypass. The boy had a VSD, and his family had traveled from South America so he could be treated by the world-famous Lillehei. With the boy's father watching the operation from the viewing area above, Barnard and his assistant opened the chest and began dissecting out the inferior vena cava. That's when the trouble started. While cutting some tissue in front of the vein, they inadvertently cut into the heart. When blood started spurting out, Barnard panicked and tried to pinch the injury closed with forceps, but the tissue tore under the force of the heart's incessant contractions. Barnard did everything he could to dampen the exsanguination, but also yelled out for the nurses to get Dr. Lillehei. As the little boy's heart stopped, and Barnard began squeezing it, begging it to work again, he remembered that the boy's father was looking down on him.

Lillehei finally arrived and effortlessly placed the boy on bypass. They fixed the VSD and then sewed up the hole Barnard and his assistant had made in the right atrium. But when they tried to come off the machine, the heart did nothing. It was dead. Lillehei left Barnard to sew up the chest, all under the eyes of the boy's devastated father.

Afterward, Barnard walked around the hospital aimlessly, not sure how to deal with what he had just done. He finally went to Lillehei's office and apologized for killing his patient.

"Look, Chris," Lillehei told him, "we've all made these mistakes that cost the lives of patients. You've made the mistake this time. The only thing you can do is to learn by your mistake. The next time you have bleeding, remember you can stop it by putting your finger in the hole. That gives you time to prepare and consolidate yourself, to get calm, and think of what you have to do . . . So tomorrow, go ahead and open the next patient's chest. We'll do the same thing. You go in and loop the venae cava and I'll wait for you."

Indeed, the next day, Dr. Lillehei stayed out of the OR until the last moment. As Barnard wrote, "Then he came in with his cocked head lamp and peered into the chest. 'Good job,' he said. 'Thank you,' I said, and thought: 'Thank you for giving me the chance to recover. Thank you for understanding how it is to lose, and how important it is to have the illusion that you can win.'" Barnard continued: "The death of the boy was tragic because it was needless—a visible mistake."

Despite their overlap at Minnesota, Shumway and Barnard were never friends, and they never would have guessed in a million years that their futures would intersect in so many interesting ways. Their personalities were so different. Shumway was relaxed, self-confident, quiet, humorous, and comfortable in his own skin; everybody loved him. Barnard was driven and brash. He knew why he was in Minnesota and would stop at nothing to achieve what he'd come to do. He was abrasive and intense, and many of his colleagues did not like him.

When the two men's training came to an end, it appeared that their paths would diverge. Barnard successfully completed all his requirements in two years. Wangensteen was most impressed, and actually asked him to stay on, but Barnard knew he had to return to South Africa.

Barnard seemed destined for greatness. He had an incredible work ethic, a fire in his belly, a belief in his own abilities, and a desire to do something exceptional. Shumway had a different story. He would have loved to stay on at Minnesota, but no offer was forthcoming. He wasn't really a researcher at that point, and he had no desire to work like the dog that Barnard was. He did remain interested in the role of hypothermia in heart surgery and thought there might be a role for localized hypothermia on the heart to protect it, and even stop it, while the body was kept alive on bypass. Eventually he took a night job running the Kolff-designed dialysis machine in San Francisco and set up a cardiac lab on dogs during the day. Not long after, Stanford Medical School moved out to Palo Alto, and Shumway went with it. This gave him more room to work in the lab and in cardiac surgery. There, he was assigned a resident by the name of Richard Lower, to help him with his lab efforts.

With dogs placed on bypass, Shumway and Lower would pour ice on the heart to see how cold they could make it and for how long and still get it pumping again. By 1958, they succeeded, with zero mortality, in restarting the heart after an hour of bypass and no heartbeat whatsoever. Imagine the possibilities this presented for the field of heart surgery. Up to that point, relatively minor surgeries on the heart (ASD and VSD repairs) were the only cases being handled by the majority of heart surgeons. With the early attempts at bypass (with a machine or cross-circulation), the heart still beat but had very little blood in it. And even these procedures, in which all surgeons did was sew up a simple hole in the heart, had a mortality rate of 50 percent. Only the best surgeons could successfully complete such cases. Yet, with topical cooling and cardiac arrest, where the heart was actually stopped, the possibility of extremely complex surgery could become a reality.

Still, the idea of heart transplantation was nowhere in Shumway's mind at this point. This was 1958. Successful kidney transplant, other than from an identical twin, was not even a reality yet.

Of his experimental surgeries with Richard Lower, Shumway said, "We would stand there for an hour with a dog supported by the oxygenator, the aorta clamped, and the heart being cooled. We were both getting bored as the dickens, so I said to Dick, 'We can take the heart out and put it in cold saline,' which we were using for cooling the heart, 'and then we can stitch it back in.'" But this they found extremely difficult to do with the rather primitive instruments and needles then available. So, they came up with the idea to get a second dog and take its heart, leaving more tissue on it so they could bolster it as they sewed it into the first dog. Thus, in this rather inglorious way, the journey toward heart transplantation began.

Shumway and his team spent the better part of the next decade researching the details of cardiac transplantation in dogs and other animals, focusing on the operation and then the postoperative care, immunosuppression, and rejection. They presented their results at surgical meetings around the country and published their findings in numerous journals. At first, their presentations attracted only a smattering of people, and the results were considered bizarre and irrelevant. But as their successes grew, so did interest among the medical community and the press. In the meantime, Shumway was growing a cardiac surgery program in humans at Stanford that boasted some of the best outcomes in the world.

Shumway was a naturally gifted surgeon, always calm, always able to find humor even in the darkest and most challenging cases. Contrast this with Barnard, a grating, driven, and in many ways tragic figure. Barnard treated many of the people around him poorly, blaming them for errors and bad outcomes. Those who worked with him recognized his brilliance and persistence, but found him annoying, demeaning, and a bit of a phony. Also, unlike Shumway, he was not a natural surgeon—he was clumsy, had a tendency to become agitated and nervous when problems arose in the OR, and suffered from constant pain and trembling in his hands from rheumatoid arthritis.

Despite all these negative attributes, he also deserves some serious praise. When he returned to South Africa from Minnesota, he was virtually alone in a country with no other legitimate heart surgeons. There had been one previous attempt at open-heart surgery, at Cape Town's Groote Schuur Hospital, but it ended in disaster. The surgeon had no previous experience using the pump on a human, and after he hooked it up, he exsanguinated the patient onto the floor of the operating room. The chief of surgery declared that there would be no further attempts at heart surgery until the return of their prodigal son from Minnesota.

Upon his return, Barnard awaited the shipment of his DeWall-Lillehei bubble oxygenator (a gift from Wangensteen) and put together a team to assist him. They practiced on dogs, simulating operations and going through all the potential disasters that could be caused by malfunction. Barnard was not fickle, or flippant about surgery, and he did not underestimate what it would take to succeed.

After months of training his team, they performed their first case together, a simple repair of a pulmonary valve stenosis (a narrowing of the valve) in a fifteen-year-old, for which Barnard figured they'd need only a short pump run. Disaster did almost strike in the OR, when a clamp came off the femoral artery in the leg and the patient nearly bled out, unnoticed, but the team managed to pull her through. Barnard sat at her bedside for days after the procedure, until he was sure she was out of the woods.

He had a few other successes in simple cases, and then expanded to more difficult ones, succeeding with complex surgeries to repair congenital defects such as transposition of the great vessels and tetralogy of Fallot. He developed his own prosthetic valves in the lab, which he used to replace diseased valves in adults. He operated on hearts with blocked blood vessels using the various mediocre techniques that existed back then. What Barnard lacked in natural technical skill and serenity in the operating room he made up for in his ability to manage patients postoperatively. He was highly detail oriented, meticulous, and had an innate ability to anticipate problems and plan for what to do about them. He would spend hours, even days, at a patient's bedside, mostly because he didn't trust anyone else to be able to deliver the kind of care he could. And he was probably a bit too honest about what drove him to do this: "I couldn't leave the patient in the hands of other people . . . You know, I've stood at patients' beds when they died, and I've been upset with everybody around me . . . It wasn't really the death of the patient—it is the ego that is hurt. I should not have had a death with this particular type of operation; I'm too good for that . . . You kill yourself for your records, but at the same time you kill yourself to save the patient." To his credit, though, Barnard enjoyed some of the best results in the world with the complex operations he took on. This continued throughout his career.

It was in those years that Barnard started thinking about the possibility of replacing the heart entirely rather than trying to repair it. In talks to medical students, he spoke about cardiac transplantation as the future of heart surgery. By 1966, inspired by the advances in kidney transplantation, he decided it was time to get educated on the advances in the discipline. He secured a training sabbatical with Dr. David Hume in Richmond, Virginia, where Hume was now running one of the major transplant programs in the world. What Barnard likely didn't know was that Hume had recently recruited Lower, Shumway's protégé, to start a cardiac transplant program. This ended up being an unexpected bonus for Barnard.

In the fall of 1966, Barnard began his mini-sabbatical at Richmond. His stated purpose was to learn about the care of kidney transplant patients, including the details of the operation and postoperative immunosuppression. He had told Hume that he was planning to start a kidney transplant program in South Africa. This wasn't a lie. Barnard saw the organ as a stepping-stone to heart transplantation, and indeed, after his time with Hume, he did one kidney transplant. It was a smashing success. His patient was still alive, with a functioning kidney, more than twenty years later.

Barnard loved his time in Richmond, and he and Hume clicked immediately. The two almost never slept. Barnard assisted Hume in the OR, went on rounds with him and his team, and immediately made his presence felt. He was attracted to Hume's tireless energy and also his swashbuckling personality.

Something else happened in Richmond. Barnard was invited into the lab where Lower and his team were practicing cardiac transplantation in dogs. Barnard stood silently watching as Lower cut out the donor heart with its atrial cuff and sewed it easily into place in the recipient's chest. Barnard was likely surprised by the ease with which Lower performed the transplant. He shouldn't have been. Lower had spent the better part of the last decade doing it in dogs. (Barnard himself had spent much less time doing so.) He was also shocked that Lower hadn't yet tried the procedure on humans.

Barnard left Richmond inspired to devote himself to cardiac transplantation and probably thought he had a good chance to be the first. He was aware of the efforts of Lower and Shumway, had read their publications since 1958, and knew that even though eight years had passed, something was holding them back from doing the surgery in humans. He likely knew what it was. Brain death hadn't been defined yet in the United States. This wasn't a problem in South Africa, where the law specifically defined death as the moment when two doctors declared a patient dead.

By this point, Barnard, Lower, and Shumway had all picked out potential recipients for heart transplantation.

Cape Town, South Africa, December 2, 1967

The day seemed like any other for the Darvall family. They had been invited for afternoon tea at a friend's house and decided to stop at a downtown bakery for a cake. Edward Darvall and his fourteen-year-old son, Keith, waited in the car while Edward's wife, Myrtle, and twenty-four-year-old daughter, Denise, ran inside to get the cake. When Myrtle and Denise left the store, they headed toward the family's car, waiting across the street. They tried to look both ways before crossing, but a large truck blocked their view of a car being driven by a drunk thirty-six-year-old salesman. The motorist was so focused on passing the truck that he barreled right through the two women. Myrtle was killed instantly, and Denise flew through the air and landed on her head. Blood poured out of her nose and mouth. When Edward saw Myrtle's lifeless body lying in the road, he knew she was gone. Denise seemed to be alive, though; at least, she was breathing.

Barnard was jolted out of an afternoon nap by the phone ringing. It was a Saturday, and he had recently returned from the hospital. He hadn't been sleeping well lately. He was consumed by thoughts of his patient Louis Washkansky, fifty-three, who had been admitted in September with severe heart failure after multiple heart attacks. Washkansky was blue, swollen, and short of breath. He was in kidney failure and had severe liver dysfunction, both of which were secondary to a massively swollen heart that was barely pumping. And yet there was still life in him. When Barnard met him a couple of months before and told him about the idea of a heart transplant, Washkansky had given a short reply: "There is nothing to think about. I'll take the chance as soon as possible." And that was it. He then ignored the rest of what Barnard said and put his head back in his book.

Washkansky loved Barnard, saw him as his savior, and called him "the man with the golden hands." Washkansky's wife did not feel the same. She did not trust the surgeon, was nervous about the idea of a transplant, and was not optimistic when Barnard gave her husband an 80 percent chance of surviving the operation. I don't know where he got that number, given that the operation had never before been done in a human being.

When Barnard got the call, he knew immediately what it meant. There was a heart available: Denise Darvall, a twenty-four-year-old white girl, blood type O, was brain dead. Barnard had promised his team that his first donor would be white—these were the days of apartheid in South Africa, and they would be pushing ethical limits already just by doing the transplant; they didn't want the negative press that would come with the very first brain-dead donor being black. But Denise Darvall was perfect. This was it. He was going to do it.

Edward Darvall, after having accepted the fact that the two women in his life were dead, for something as stupid as picking up a cake, consented to the donation. "If you can't save my daughter, you must try to save this man," he said. Maybe the knowledge that his daughter would be the first donor for a human heart transplant was a comfort to him.

Barnard knew he would be criticized for removing the young woman's heart, but he also knew it was the right thing to do:

Denise Darvall had entered a no-man's land between life and death—an area created by modern science and medicine. She was being held there by drug stimulants, blood transfusions, and, most important, artificial breathing provided by the automatic ventilator. How long it would take her to cross over to total death depended mainly upon how long we continued to run the ventilator. A flip of the switch, turning it off, would result in immediate cessation of breathing. Her heart would continue to beat for three, four, maybe five minutes—and then stop.

At that point we would have the three criteria that doctors have used for centuries to determine death: no heart beat, no respiration, and no brain function. Denise Darvall, who had been med-

ically dead, would then be legally dead. We could consign her for burial—or, as we intended, open her chest and remove her heart. On the other hand, if we restarted the ventilator immediately, and at the same instant gave her heart an electric shock, we could in all likelihood set again in motion the twilight existence we had just terminated. From being legally dead, the patient would be returned to the same no-man's land she had just left—and where she could continue to exist for an indeterminate length of time as a biological vegetable.

At 2:20 a.m. Denise's ventilator was turned off. She was hooked up to catheters so that the bypass pump could be turned on with a flick of the switch, and then the team sat and waited. As she was already brain dead, she would not take a breath on her own. With no respiration, no oxygen would enter her lungs, diffuse across the capillary membranes into the bloodstream, bind with the hemoglobin in her blood, and be delivered to her organs, including her heart. Cells in these organs would start dying, and the organs would stop functioning. At some point, the heart would stop beating. How long would that take? Minutes, most likely; agonizing minutes, with every tick of the second hand representing more dead heart muscle.

Barnard was in his right *not* to wait. South African law would have supported him if he had clamped the vessels in and out of the heart and cut the organ out. So, what should he do? This is what he reported: "So we waited, while the heart struggled on—five, ten, fifteen minutes. Finally, it began to go into the last phases, its wild peaks slowly sinking into exhausted rolls that became longer and longer until it finally revealed itself in a straight green line across the screen—death. 'Now?' asked Marius. 'No,' I said. 'Let's make sure there is no heart beat coming back.'"

That wait must have been excruciating for Barnard and his brother Marius, who was assisting him in the procurement of the heart. At least, it would have been excruciating if it in fact had happened. Roughly forty years later, after Christiaan Barnard was dead, Marius Barnard told the writer Donald McRae in an interview that, with Barnard's consent, he, Marius, had injected a huge slug of potassium to stop the heart right after they shut off Denise's ventilator. Then they opened her chest, placed catheters carefully in the right atrium for drainage and a return catheter in the aorta, and flipped the switch on the bypass machine, making the heart return to its pink color. They then cooled Denise down, and Barnard went next door to check on Washkansky.

Barnard's team had already opened Washkansky's chest, revealing the massively enlarged beast of a heart that was (barely) keeping him alive. Barnard and his team placed Washkansky on bypass, and nearly lost him in the process from a technical error with the pump. Fortunately, they were able to work through this, but it reminded Barnard how dangerous the bypass pump could be.

Barnard returned to Denise's room and began cutting her heart out. While he may not have had the same decade of experience on dogs that Shumway and the others had, he knew what he needed to do. He cut around the heart carefully, being sure to cut the vessels at angles, so they could join nicely with Washkansky's larger vessels. He then returned to Washkansky's room carrying a metal bowl with this incredible gift in it. There, he cut out Washkansky's massive, useless heart, making sure to leave a lid of atria so he would be able to connect up all the vessels. Barnard's own heart was pounding as he did this. He knew there was no room for error. Once he removed Washkansky's heart—which, amazingly, was still trembling when he put it into a different metal bowl—he took Denise's small, beautiful heart in his hands.

He later wrote, "For a moment, I stared at it, wondering how it would ever work. It seemed too small and insignificant—too tiny to ever handle all the demands that would be put upon it. The heart of a woman is 20 percent smaller than a man's, and the heart of Washkansky had created a cavity twice the normal size. All alone, in so much space, the little heart looked much too small—and very lonely."

Barnard then got to work—first sewing the left atrium, then the right atrium. It looked perfect. Then he and his team moved on to the pulmonary artery. Because he had cut it at a branch patch, meaning he made a larger orifice by cutting it where it branched into two arteries, it matched perfectly with Washkansky's. The donor and recipient vessels came together perfectly. Now the aorta.

Barnard trimmed Denise's aorta at an even more severe angle, to increase the size of the lumen (or orifice), trying to make it match the large aorta hanging out of Washkansky's chest. As he did this, he told his team to start warming Washkansky. At 5:15 they started sewing the aorta, and at 5:34 they were done. The heart was in—but it was blue. Would it work?

Barnard loosened the snares on the cavae and let blood flow through the heart. The heart swelled with warm blood. It started to fibrillate. Barnard and his team watched, hoping that the heart would develop a coordinated rhythm and worrying that it wouldn't. Barnard had done this on dogs at least fifty times, but it hadn't always worked. He called for the paddles and gave the heart a shock, twenty joules. It froze for a second, and then, slowly, started to contract. The contractions started in the atria, and then the ventricles.

Slowly, they picked up pace, until they reached 120 beats per minute. They had a heartbeat, but would it be strong enough to support this big man?

Barnard prepared to take Washkansky off the bypass machine. When everything was ready, he delivered the command "Pump off." But Washkansky's blood pressure started to drop. One of the nurses called out the numbers: "Eighty-five . . . Eighty . . . Seventy-five . . ." The heart looked distended and unhappy. "Sixty-five." Barnard had the team put the pump back on. They gave Washkansky some IV drugs, corrected some electrolytes, and checked his temperature. They tried again—and the same thing happened.

Barnard tried to sound confident, but inside he was dying. "I was horrified," he would later admit. But he persisted, and on the third try, at exactly 6:13 in the morning, it worked. They turned the machine off for the last time.

Barnard scrubbed out and went into the tearoom. His brother Marius joined him. Washkansky's new heart was holding at a steady 120 beats per minute. Barnard took his own pulse: it was 140. He stayed around the hospital for the next few hours, making sure Washkansky had been settled into the ICU and was stable. He finally made his way home at around noon, and shortly after that, the world exploded.

It started with a short report on the local radio, about a heart transplant that had just taken place—Barnard's name wasn't mentioned. Then, over the next hour, the report expanded, and that's when the phone started to ring. Barnard fielded calls from all over the world. One of the earliest was from a reporter in London whose first question was whether the patients were black.

That night Barnard couldn't sleep, and he made his way back to the hospital. He spent the next eighteen days essentially at Washkansky's bedside. Over those eighteen days, Washkansky recovered, was able to have his breathing tube removed, was talking, and was wheeling around the hospital and improving. Unfortunately, after a couple of weeks, he started having fevers. He developed a severe postoperative pneumonia and, despite a beautifully working heart, succumbed to infection, likely because of all the immunosuppression he was receiving.

Barnard was crushed by Washkansky's death, but the rest of the world didn't seem to care. Barnard was an international superstar. He was about to begin a tour around the United States that would include an afternoon with the president, appearances on *Face the Nation* and *Today*, and cover stories in *Time*, *Life*, and *Newsweek*.

The next transplant Barnard performed was on January 2, 1968. The patient was Philip Blaiberg, a fifty-eight-year-old retired dentist with a failing heart. After Washkansky died, Barnard had met with Blaiberg and his wife to tell them the news and see if Philip still wanted to proceed with his transplant. "Professor, I want to be a well man, and if I'm not well, I'd rather be dead." Blaiberg's donor was a twenty-four-year-old black man who had suffered a sudden brain bleed while on a beach. He was declared brain dead, and he and Blaiberg were brought to the operating room for transplantation.

Barnard again encountered trouble with the bypass pump, but was able to work through it. Maybe even more amazing, just as he was getting ready to take the transplanted heart off the pump, the power in the hospital went out. They were standing in the dark, with no electrical power going to the bypass pump. Barnard quickly had the team remove the venous tubing and hand-crank the bypass pump and ordered rapid rewarming. As Barnard stopped the pump, the heart fibrillated (quivered), slowly coming to life. When the lights came on, the rhythmic beating had already commenced.

Blaiberg was extubated by post-op day one and was discharged after a ten-week stay. He went on to live an astounding nineteen months; photos of him during that time show him enjoying the beach and other activities. With Washkansky, Barnard had proved the operation could be done in humans, and with Blaiberg, he proved that heart transplantation would eventually become a viable option for patients with irreparable heart disease.

The Americans Get Their Chance

On January 6, 1968, in California, Shumway got his first chance at a transplant. The donor was a forty-three-year-old woman who had suffered a brain bleed. The recipient was a fifty-four-year-old steelworker who had severe heart disease. The surgery itself went well, but the recipient suffered virtually every complication known to man. Shumway did everything he could, including multiple reoperations, but the recipient died on post-op day fourteen.

Dick Lower finally got in the game on May 25, 1968. The donor was Bruce Tucker, a fifty-six-year-old black man who worked at an egg-packing plant. He had been drinking that evening after work with a friend, and when he got up to stumble home, he fell and hit his head on the pavement. He was brought to the hospital at 6:00 p.m., unconscious with a devastating head injury. The neurosurgeons took him to the OR later that night, but by the next morning, the examining physician wrote a note that read, "[T]he prognosis for recovery is nil. Death is imminent." By 1:00 p.m., after an EEG failed to show any cerebral brain activity, the staff neurologist agreed with that assessment.

Hume and Lower were notified. Hume contacted the police and

asked them to search for the dead man's next of kin. It is hard to know how extensively they searched, but at 2:30 p.m., they let Hume know the man's family could not be located. Hume obtained permission from the state medical examiner, and with that, he urged Lower to proceed. Tucker's respirator was disconnected at 3:30 p.m., and minutes later, after Tucker's heart ceased beating, Lower began splitting his chest. His heart and kidneys were removed, and Lower proceeded with his first heart transplant in a human. Lower was probably the most experienced heart transplant surgeon in the world by this point, having performed this surgery in hundreds of dogs. As expected, the surgery was a total success, and the recipient, Joseph Klett, enjoyed this gift from Bruce Tucker for a whole week until he died from rejection.

A few days after the transplant, Bruce Tucker's brothers were finally located and came to the morgue to claim their brother's body. It was only then that they discovered that his heart and kidneys had been removed for transplant. This was devastating for the Tucker family. Not only had they just found out that their brother had died alone in the hospital, but his organs had been harvested from his body without the family's knowledge or consent. And he was taken to the operating room while his heart was still beating. Sure, the neurologist had considered Tucker brain dead, but this was three months before the diagnosis of brain death had been defined in the American literature and more than a decade before brain death became legally synonymous with death. I can only imagine the effect this had on the Tucker family. After all, this was Richmond, Virginia, in the 1960s, not the most hospitable location for blacks in the United States.

No criminal charges were filed, but the Tuckers did bring a lawsuit against Hume and Lower. It was prosecuted in civil court in a seven-day trial. The Tuckers were represented by Douglas Wilder, later a state senator and, in 1990, governor of Virginia, the first African American to be elected governor in the United States. Wilder argued that "the transplant team engaged in a systematic and nefarious scheme to use Bruce Tucker's heart and hastened his death by shutting off the mechanical support systems." He also hammered home the point that Tucker was labeled "unclaimed dead," and that he belonged to the "faceless black masses of society."

Early in the trial, the judge instructed the jurors to stick to the "legal concept of death and reject the defense's attempt to employ a medical concept of neurological death in establishing a rule of law." This may have signaled the death knell for Hume and Lower's case, but they were helped by the fact that, since the transplant, a major article defining brain death had been published in *The Journal of the American Medical Association* and was generally supported among the thought leaders of the time. Hume and Lower's team defended the concept that Tucker was truly dead when his brain died, and because of that, the removal of his heart was not what had killed him. Their expert witness list included Dr. Joseph Fletcher, a well-respected professor and bioethicist. Fletcher convincingly summarized Tucker's state at the time of his organ donation as follows: "When cerebral function is lost, nothing remains but biological phenomena at best. The patient is gone even if his body remains and even if some of its vital functions continue. He may be, technically, 'alive.' But he is no longer human. He is, as a human being, undoubtedly dead."

After seventy-seven minutes of deliberation, the jury returned with a verdict of "not liable." The removal of Tucker's heart for organ donation had not caused or accelerated his death. After the trial, Hume spoke confidently of their victory for the press: "This simply brings the law in line with medical opinion . . . I think this is an issue that had to be decided, and I think it will have an influence on the medical community for a long time to come." He was right about that.

Regardless of the outcome of this trial, a great disservice was done to Bruce Tucker and his family. Hume and Lower were so motivated to get into the heart transplant arena that they proceeded with the transplant without thinking about the effect it might have on the Tuckers. This goes against my very core belief about organ transplantation: that the donors (and their families) are our patients, too. They are the heroes, the ones who make it all happen, and they also benefit from the process. Hume and Lower focused solely on their recipient, and thus deprived the Tuckers of that benefit. The fact that Tucker may have been dead (or at least not alive) does not mean he and his family did not still own his body, even if his soul was gone. Hume and Lower disrespected that.

The progression of heart transplantation after Barnard's first was similar to the explosion in flight after the Wright brothers proved its potential in 1903. Before their success, few people could imagine it being done, but within a year, more than 100 flights had been completed by numerous pilots. So, too, with heart transplantation. In 1968, 101 transplants were performed in 26 countries. Every major hospital wanted a program, and prominent heart surgeons around the world jumped on the bandwagon.

One great example was Denton Cooley in Houston, possibly one of the most technically gifted heart surgeons in the world. When he heard about Barnard's first successful transplant, he sent the following cable: "Congratulations on your first transplant, Chris. I will be reporting on my first hundred soon." He wasn't kidding. A few months later, in April 1968, he did his first, which took him thirty-five minutes to sew in. His patient made it out of the hospital and back to work (although he died after seven months). Cooley completed seventeen by the end of 1968, but only two of the patients lived longer than six months. Cooley was a master surgeon, but he was not a transplant surgeon. Few heart surgeons were then. By mid-1969, he shut down his heart transplant program, as did the head of almost every other program in the country. Due to the miserable outcomes, physicians were loath to refer patients as donors and donor consent rates were low.

Roy Calne described the phenomenon this way:

In the eyes of the media, the drama of the operation seems to be more important than the long-term well-being of the patient, which is what transplantation is really about . . . The surgical results were usually satisfactory but most of the cardiac surgeons had no experience of transplantation immunology or immunosuppression and the infrastructure needed to prevent or control destructive rejection. Virtually all of these poor patients perished, having satisfied the macho aspirations of their surgeons. This series of failures had an extremely bad effect on the image of transplantation and resulted in a self-imposed moratorium on heart transplantation, except in a few centers where the procedure could be done with the appropriate infrastructure.

In 1971, just seventeen heart transplants were performed in the world.

In the end, outcomes wouldn't really improve until the discovery of cyclosporine. As was the case for all transplants, the 1970s were the tough years, when the believers toiled on, despite mediocre results, until better immunosuppression became available. And the main believer was Shumway. A true transplant surgeon, he carried the torch through the wasteland of the '70s and into eventual success in the '80s. Without a doubt, although Christiaan Barnard was the first, Norman Shumway was clearly the father of cardiac transplantation.

KIDNEY TRANSPLANT LED to a Nobel Prize, with a handful of famous surgeons involved and the most prominent hospital in the world at center stage; heart transplant generated international excitement and chaos; and liver transplant (which I'll discuss in chapter 9) was championed by the biggest name in transplant, perhaps the most recognizable personality in our field, Thomas Starzl. But does anyone know who performed the first lung transplant? For most, the answer is no.

Actually, the man who performed the first lung transplant in humans was the same one who did the first heart transplant in humans. No, not Christiaan Barnard. It was James Hardy, at the University of Mississippi Medical Center, in Jackson. While Barnard performed the first heart transplant between two humans, three years earlier, in 1964, Hardy had performed the first heart transplant between a chimp and a human. He took an incredibly ill sixty-eight-year-old man with a failing heart and dying limbs, truly hours from death, and sewed a chimpanzee heart into him. The surgery itself went well, and the heart did start beating immediately, but Hardy couldn't get the patient off the bypass pump. The tiny chimp heart just couldn't beat strongly enough to sustain the recipient's blood pressure. Ultimately, it proved too small for the patient, and he died in the operating room. Hardy was demolished in the press and the transplant community.

This horrible day may have overshadowed a major accomplishment for Hardy of just seven months prior. The year was 1963. Hardy had performed almost four hundred lung transplants in dogs. He had the operation down, but given the quality of immunosuppressive therapies at the time, the vast majority of recipients were dead by four weeks post-op. Nevertheless, Hardy thought it was time to try the procedure on a human.

He met the perfect recipient on April 15, 1963. John Richard Russell, a fifty-eight-year-old white male, was extremely ill and just what Hardy had been looking for, absolutely perfect, presenting no red flags whatsoever—except maybe just a couple: He was a heavy smoker and had a massive tumor completely obstructing his left main-stem bronchus, with pus around it, and a diseased lung on the other side. He was also in chronic renal failure and was approaching dialysis. Oh, and one other little thing: he had been convicted of murdering a fourteen-year-old boy in 1957 and was serving a life sentence. In his defense, Russell had maintained that the murder was accidental, but he had still been convicted.

At the very least, it was clear that Russell was dying. Hardy approached him about the possibility of a lung transplant. When writing about the case after its conclusion, Hardy clarified how the decision was made: "Although the patient was serving a life sentence for a capital offense, there was no discussion with him regarding the possibility of a change in his prison sentence. However, authorities of the state government were contacted privately, and they indicated that a very favorable attitude might be adopted if the patient were to contribute to human progress in this way."

Six weeks later, a man was brought into the hospital with a massive intracranial hemorrhage. He was undergoing CPR, and when it became clear that he would not survive, his family was asked to consent to organ donation. Hardy removed Russell's left lung and then sewed in the donor's left lung, hooking up the blood supply and the airway just as he had hundreds of times in dogs in the lab. The surgery went smoothly, and the lung functioned right away. Unfortunately, after the operation, Russell's kidneys failed. While he was dying, a major press release announced his pardon by Governor Ross Barnett of Mississippi, who thanked him for his courage in trying to help mankind. At least he died a free man, nineteen days after the world's first lung transplant.

Hardy's poor outcome with a prisoner, followed by his poor reception after the chimpanzee transplant, convinced him to lie low and let others battle through all the bad outcomes that would accompany transplantation in the 1970s. And so, the discipline of lung transplant limped along almost imperceptibly, with no real success for almost twenty years.

Between 1963 and 1981, roughly forty single lung transplants were performed, with virtually all recipients dying from infection or technical complications. The biggest challenge was the connection to the airway, which would invariably leak into the patient's chest, causing massive infection and death. This likely was due to inadequate immunosuppression, which led to early rejection. In all those years, there was only one success: a twenty-three-year-old patient in Belgium who received a single lung in 1968 for the diagnosis of advanced pulmonary silicosis. (He was a sandblaster.) Despite some early rejection episodes, he survived for ten months before dying of pneumonia. His autopsy revealed a healthy-looking lung transplant with no evidence of rejection.

While Hardy falls into the long list of surgeons who wanted to be first, Norm Shumway wanted to be right. By the late 1970s, having established himself as the premier heart transplant surgeon in the world, he was bothered by the subset of patients he couldn't help: those who needed a heart transplant but also had severe lung disease. Many of these patients had congenital heart disease that had gone untreated and, over time, had damaged the lungs. Why couldn't he take the heart *and* lungs en bloc and sew them in together? But first, he needed to know that he had a reasonable chance for success.

The timing couldn't have been better for Bruce Reitz. While an undergraduate at Stanford, Reitz became interested in the heart when he studied the immunologic reactions of the heart as a physiology major. He worked in Shumway's lab as a medical student (eighteen months after Shumway's team performed the first heart transplant in man) and returned there after his residency and cardiothoracic fellowship at Stanford. Shumway asked Reitz to focus on combining heart transplant with bilateral lung transplant, and Reitz began by performing autotransplants in monkeys, removing their hearts and lungs while they were on bypass and then sewing them back in. In this way, Reitz and his team could perfect the technique without having to deal with rejection. Once they had done so, they moved on to allotransplant, removing the same organs from donor monkeys and placing them into recipients. But the outcomes were not adequate. Immunosuppressive therapies just weren't good enough yet.

Things changed in the summer of 1978, when Reitz and his team were able to obtain cyclosporine for their research. This new wonder drug changed everything. With cyclosporine part of the antirejection regimen in their monkeys, the tracheal (airway) anastomosis healed and the monkeys survived.

By the fall of 1980, Reitz and Shumway thought their results in the lab good enough for them to consider a heart-lung transplant in a human. Enter Mary Gohlke, a forty-five-year-old woman whose pulmonary hypertension had led to a failing heart and lungs. Gohlke knew she couldn't wait much longer, and was able to reach her senator in Arizona, who pushed the FDA to give Stanford approval to use cyclosporine in combined heart-lung transplants. On March 9, 1981, Reitz and Shumway removed Gohlke's heart and lungs. In Reitz's own words, "The appearance of Mary Gohlke's totally empty chest was indeed a dramatic moment. I wondered, 'Is this really going to work out?' But the implantation went smoothly, the heart resuscitated quickly, and lung function was adequate immediately." Gohlke went on to live five more years, and when she died, her organs were still functioning, with no rejection noted during her autopsy.

TWO YEARS AFTER Gohlke's transplant, in 1983, the first successful lung transplant alone was performed by Joel Cooper at the University of Toronto. The patient was Tom Hall, a fifty-eight-year-old with pulmonary fibrosis (scarring of the lung tissue). Cooper had previously taken part in the forty-fourth failed attempt, in the late 1970s, and realized that doing more in humans at that point would have been fruitless. So, in the tradition of all great pioneers, he went back to the lab. Knowing that the factor killing all their patients was the airway's inability to heal, he focused on wound healing, and found that high-dose steroids were most to blame. Like Shumway and Reitz, he realized he'd need something different. By the time cyclosporine was available for use in humans, Cooper had perfected a novel technique for sewing the airway and bolstering it with omentum, vascularized tissue found in the belly. Although he now thought it time to resume human transplantation, he was not sure what kind of outcome he could expect. He recalled his conversation with Tom Hall before the surgery: "I said, 'Tom, there have been about 44 attempts thus far and no one has survived. Are you sure you want to go ahead with it?' He said, 'I am grateful to be number 45.'" Hall lived more than six years, truly returning to a normal life.

Cooper went on to conduct the first successful double lung transplant in 1986 as well. Lung transplant (like pancreas transplant, discussed in chapter 8) has some particular challenges. Unlike with heart, liver, and kidney transplantation, which became relatively safe after the introduction of cyclosporine, lung patients, even as recently as the early 1990s, had only a fifty-fifty shot at leaving the hospital alive. Infection has always been one major hurdle. In 1990, 290 cases were reported, with 65 percent survival at one year and 54 percent at two. But due to the commitment of pioneers such as Reitz and Cooper, the likelihood of leaving the hospital today is close to 97 percent. The current one-year survival rate is roughly 80 percent, with five-year survival just better than 50 percent. For unclear reasons, lungs seem to be susceptible to a unique type of chronic rejection that limits long-term outcomes. Numerous investigators are trying to understand this better. It is hoped that outcomes will continue to improve.

| 8 |

Sympathy for the Pancreas

Curing Diabetes

When a child is diagnosed with type 1 diabetes, an entire family is diagnosed with type 1 diabetes.

—TYPE1MOMS (WWW.TYPE1MOMS.ORG)

Insulin is not a cure for diabetes; it is a treatment. It enables the diabetic to burn sufficient carbohydrates so that proteins and fats may be added to the diet in sufficient quantities to provide energy for the economic burdens of life.

—FREDERICK BANTING, "DIABETES AND INSULIN," NOBEL LECTURE, SEPTEMBER 15, 1925

There's an old saying in surgery. "Eat when you can, sleep when you can, and don't mess with the pancreas." I think every surgeon has at least one experience related to a leaking pancreas, and it likely involves multiple drains, an open wound, and a miserable patient.

Yet there is a certain amount of celebrity associated with surgeons who operate on this organ. The Whipple, in which the head of the pancreas is removed because of cancer or benign lesions, is one of the most fabled operations, and those who do it are considered high priests of abdominal surgery. The irony is that along with the complexity of the operation and the high rate of complications (two things that frighten off the most courageous surgeons), the operation doesn't even work very well. While some surgeons report a five-year survival rate of 20 percent, most patients who live are found not to have had cancer in the first place, but rather, inflammation or a precancerous lesion. I understand that patients are trying their best to survive, but with the most famous surgical operation, the one that is like slaying the dragon, patients usually end up getting eaten by the dragon anyway.

In organ transplant, we don't just mess with the pancreas; we roll it up, squeeze it out, and transplant the whole thing for patients afflicted with diabetes.

As I finished my residency in Chicago, I wanted to continue my training at a top-notch transplant center. I was interested in learning liver transplantation (the Super Bowl of the abdomen), but I also wanted to have the opportunity to mess with the pancreas. I got that chance at the University of Wisconsin—and it didn't disappoint.

In my first two months in Madison, I performed sixteen pancreas transplants. Most of them were with Hans Sollinger, perhaps the premier pancreas transplant surgeon in the United States. Hans started in pancreas transplant in the early 1980s, when outcomes were dismal, mostly because the connection between the pancreas and the bowel kept falling apart. Across the country, programs were giving up on the operation. Some surgeons were letting the pancreatic duct drain freely into the abdomen or were bringing it out through the skin and collecting the fluids in a bag. Around that time, Dr. Folkert Belzer, the chairman of surgery at the University of Wisconsin, told Hans he'd better figure out how to make pancreas transplant work or he'd shut the program down. In a fit of frustration, Sollinger yelled out, "How 'bout I sew the pancreas to the fucking bladder!" Though Sollinger said this in desperation, both he and Belzer realized it was a good idea, and that's what saved the pancreas program at Madison.

This was in 1983, and for the next fifteen years, Sollinger's idea (which he first tested in dogs) became the primary technique used by surgeons across the globe. Around the mid-1990s, after documenting that patients were suffering long-term urinary tract complications due to the drainage of pancreatic juices (and given the improved immunosuppression, which allowed for better healing), Sollinger switched back to drainage into the bowels. I still see this as one of his finest moments—even though he had popularized bladder drainage, which played a huge role in the success of the entire discipline, he was one of the first to publish and lecture on its problems, and he encouraged surgeons to switch back to bowel drainage. That took guts!

I'M NOT ONE to judge the patients I evaluate, but I accept that many diseases are caused at least in part by the choices we make. If we got everyone to stop smoking and abusing alcohol and drugs, to eat better, and to clean up the environment, we doctors probably would have very little to do.

None of this applies to patients with type 1 diabetes of course. People develop this horrible disease through no fault of their own, typically at a young age, when they are supposed to be enjoying the world around them without thinking about consequences. And the consequences of diabetes are brutal: amputated limbs, blindness, heart disease, kidney failure, and perhaps most cruelly, impotence. Patients diagnosed with type 1 diabetes are usually kids, tend to be thin, and in general look quite healthy. They are often active, play sports, and do well in school. What we don't see is how different their lives are from those of other kids. Anytime they eat, they have to think about how much insulin they should take. Their fingers become hardened from all the needle pricks. In many cases, the disease plays a major role in defining their personalities. When all the other kids at a birthday party run over to eat cake and ice cream, the kids with type 1 diabetes run over to their parents to ask if they can, knowing that they will be told no. Kids with type 1 are awakened a couple of times every night so their blood sugar level can be checked, to make sure they don't dip too low or soar too high while they're sleeping. These kids are forced to be organized, prepared, and detail-oriented. When you evaluate them in the clinic, they always ask dozens of questions, write everything down, and think through every detail (or their parents do).

While in general it has been hard to show a significant benefit to pancreas transplant in terms of survival (unlike with a kidney transplant, which clearly prolongs life), there are no more thankful patients than the recipients of pancreas transplants. For the first time in their lives, these patients are not defined by their blood sugar. They can eat what they want, they can sleep through the night without worrying they won't wake up, and they don't have to call themselves diabetic anymore.

I recently evaluated Mary J. for a pancreas transplant. A pleasant forty-five-year-old woman, she came with her four kids in tow, ages two to fourteen. She looked totally healthy, and for a minute I wasn't sure what she was doing there. In general, we don't perform pancreas transplants for type 1 diabetes until patients have developed kidney failure. That's because the antirejection medications have so many side effects that the benefit of getting off insulin may be outweighed by the disadvantages of taking immunosuppressive drugs. Still, we will consider doing a pancreas transplant alone for patients who have such brittle diabetes (a severe form that involves unpredictable swings in blood sugar level and difficulty getting sugars under control) that they develop hypoglycemic unawareness (a complication in which the patient is unaware of dangerous drops in her blood sugar level, which can lead to seizures or death).

Mary is one of those patients. She has episodes, pretty much every other day, in which she will suddenly black out. She could be at work, at the grocery store, or asleep. Once, she blacked out while driving her kids in the car. She has had to give up her job and needs help taking care of her children. Her diabetes is so hard to control that her blood sugar level either flies off the charts or bottoms out, becoming low enough to cause seizures. But unlike most diabetics, Mary can't tell when her blood sugar is getting low. There is no warning, no sweating or light-headedness. Her glucose level rises and drops so precipitously that an insulin pump hasn't helped, and her life has become unlivable.

Yet all that will be fixed with a pancreas transplant. (Perhaps there is one other option for her: A recent patient of ours had a dog that could detect when her blood glucose level got too low based on her smell and taste when he licked her skin. If her level got out of whack while she was asleep, he would wake her up so she could take some sugar or insulin. We gave her a transplant, which worked great, but the dog lost his job.)

IN THE EARLY 1900s, there was no successful treatment for type 1 diabetes. Children who developed this disease were placed on starvation diets until sugar could no longer be detected in their urine, with the hope that someone would figure out how to treat them before they died of starvation. By the turn of the twentieth century the medical community recognized that the hormone responsible for handling sugar in the blood (eventually to be called insulin) came from the pancreas, though numerous researchers failed in their attempts to isolate this protein. Finally, in January 1922, insulin was isolated and injected into a patient by the most unlikely of investigators, Frederick Banting, a surgeon with virtually no surgical practice or research background. His persistence, or stubbornness (likely compounded by a case of PTSD from his service in World War I), together with the help of a couple of real scientists, J. J. R. Macleod and James Bertram Collip, and a tireless medical student, Charles Best, at the University of Toronto, led to the isolation of insulin and its use as a life-saving treatment for thousands of children diagnosed with this disease. In the 1920s and '30s, there was hope that insulin would return diabetic patients to normal lives (other than having to inject insulin multiple times a day). By the mid-twentieth century, patients with type 1 diabetes were living into their forties and fifties, but were now facing the horrible complications just mentioned, including kidney failure. Although the discovery of insulin prolonged the lives of these patients, the difficulty in measuring blood sugar and knowing when and how much insulin to use and the inability to tightly regulate sugars throughout the day and night were still major problems. It seemed possible that transplanting the whole pancreas could solve all this.

There were some attempts in the 1950s to transplant pancreata in dogs, but the vast majority of transplants led to devastating complications from the leaking of highly destructive pancreatic enzymes. Various techniques were employed to prevent this, ranging from tying off the pancreatic duct (which led to scarring that ruined the organ) to bringing the duct out through the skin and collecting the enzymes in a bag. Some investigators tried radiating the organ to destroy the parenchyma, the tissue responsible for creating the digestive juices (hoping it wouldn't also destroy the insulin-secreting islet cells), but that failed as well. Others included a cuff of duodenum, the bowel that collects the drainage of pancreatic juices as they exit the pancreatic duct, in the anastomosis rather than just using the pancreatic duct. Initially, the bowel would be attached to an adjacent loop of small bowel in the recipient, allowing these juices to flow through the intestine (which is where they go normally).

By the late 1960s, some experiments in dogs were showing a survival rate of half a year or more, good enough for a few investigators to attempt the procedure in humans. On December 17, 1966, a team at the University of Minnesota that included Drs. William Kelly and Richard Lillehei (the younger brother of C. Walt Lillehei) performed the first pancreas transplant in a human, a twenty-eight-year-old woman, along with a kidney transplant to cure the renal disease caused by her diabetes. In this operation they opted not to include the donor duodenum but, rather, ligated the pancreatic duct. The blood vessels to the pancreas were anastomosed to the iliac vessels, a technique similar to that used for kidney transplantation. The pancreas worked right away, allowing the patient to come off insulin and have normal blood sugar levels regardless of what she ate. Sadly, she died after two months, having suffered rejection of the organ and, ultimately, infection.

Heartened by the mild success of this transplant, the team

performed thirteen more pancreas transplants over the next seven years. Nine of them included a kidney as well, and the majority was transplanted with a cuff of the duodenum sewn onto adjacent small bowel. Although one graft survived for a year, the rest were lost in a much shorter time, with the patients dying secondary to infection or renal failure. It is fair to say pancreatic transplantation was not accepted as a viable option in those days, given the success of insulin and the reasonable-ish outcomes of transplanting kidneys alone in diabetic patients. A few researchers soldiered on, though, led primarily by the Minnesota team. By 1977, fifty-seven pancreas transplants had been performed with only one long-term survivor. Many of these grafts were lost because of leaking at the site where the duodenum was sewn to the bowel, causing surgeons to remove this cuff of duodenum and transplant the pancreas alone. Outcomes for pancreas transplant didn't really start to improve until the advent of cyclosporine in the 1980s, which dramatically reduced organ rejection.

Pancreatic transplant is unique among organ transplants. Unlike the heart, liver, and kidney, all of which are necessary for the patient's survival, the pancreas primarily provides a quality-of-life benefit. The patients could otherwise survive on insulin, albeit as slaves to their own blood sugar. After witnessing a few horrible complications after pancreas transplant (requiring multiple returns to the OR, leaking organs and drains, and open wounds), I wondered if it was worth it. Wouldn't it be better to use an insulin pump and get a kidney if needed?

I thought this, that is, until I got to know some pancreas recipients and heard their stories. While there are many successful ones, which I could share, sometimes the most complicated cases are the most revealing.

El Diablo (or Disconnected Pancreatic Duct Syndrome)

I had my first encounter with El Diablo about eight years ago. I was on my way to my lab when my phone rang. On the screen was the name of one of my partners, Janet.

"Hey, I'm getting killed today," she said. "Think you can do a quick ex lap? It's some previous transplant with free air, probably a colon perf or duodenal ulcer."

I pushed aside the multiple excuses that entered my brain, ranging from the lab work I wanted to do to a dread illness I was developing. "Sure, no problem."

Twelve hours later, I was still in the OR. Quick, my ass. The patient was many years out from a combined kidney-pancreas transplant. He had recently been treated for rejection, and had recovered from this, with the pancreas still functioning, although not as well as before. His blood sugars were borderline, and he was close to needing insulin. His kidney wasn't working that well, either—he still wasn't on dialysis, but he was not far away.

Earlier in the week "JB" was at home working on his roof when he fell off the ladder and hit his belly. A few days passed, and the pain that had started out mild had increased in severity until he finally made his way to the hospital. A CT showed free air, indicating a perforation in a portion of his bowel, usually a clear indication of the need for emergency surgery.

When we got in, we found adhesions everywhere—that is, scar tissue most likely formed from his previous surgery that had caused all of JB's bowel to stick together. We spent hours dissecting out the various tissues, taking care not to injure any of the bowel as we went along. After many hours of this, we deduced that there was no hole in the stomach, duodenum, or colon. When we got to the transplanted pancreas, however, we found a hole right where the duodenum had been sewn onto the small bowel during his years-old kidney-pancreas transplant. The hole was leaking pancreatic and small bowel contents (succus—we generally use the term *succus*, or *succus entericus*, to describe small bowel contents) everywhere. Given the amount of spillage and how sick the patient was, I knew I had to take out JB's pancreas. It wasn't working that well anyway, and clearly if I left it in, it was going to kill him. I would need to resect the portion of bowel onto which the duodenal cuff had been sewn and then hook that bowel back together. That should be easy. I would also need to find the artery and vein that supplied the organ, which had been sewn onto the iliac vessels going down to his right leg. Finally, I would need to scoop out the body and tail of the pancreas.

Digging out the blood vessels proved to be nearly impossible. Everything was so stuck that it was like carving through a piece of hard marble. That's the thing about scar tissue. You never know how thick it will be, how difficult to take down, until you get in there. Usually transplant patients have less scar tissue than other patients who have undergone surgery—the immunosuppression actually suppresses its development. But not always. Eventually I carved my way across the vessels and divided them.

After reconnecting his bowel, I tried to peel out the pancreas. It was so stuck and surrounded by so much inflammation that, to be honest, I couldn't tell what was pancreas and what was surrounding tissue that needed to stay in there. Once I had removed a handful of yellow tissue, I placed a drain and closed up. After talking to JB's family, I made my way home at around midnight.

When I spoke to JB the next day, he hadn't been told yet that I had removed his pancreas. He had been too sedated after surgery, so I figured I'd let him know when I saw him on my rounds. He was devastated. I told him we'd had no choice, that we'd had to take the organ out to save his life.

"Well, how long until I can get another one?" he asked.

I told him it was too early to talk about that. "Let's just focus on getting through this for now." Before I left, I looked at the drain I'd put in him. The bag had pinkish fluid in it, and I noticed that the output was pretty high—almost half a liter. *Hmm. Should dry up . . .*

The next morning at around 6:00, I flipped through JB's chart. His labs looked okay, but I again noticed that he had drained almost another liter of fluid overnight. Could I have left some of that pancreas in there? I texted my fellow, John, and asked him to have the fluid tested for the presence of amylase, an enzyme made by the pancreas. By 10:00 a.m. I saw the result: well over 1,000 (way too high; the level in normal body fluid should be under 100). There clearly was some pancreas still in there that was leaking amylase. I figured the leak would dry up; after all, the damn organ didn't have any blood supply, it had to die. But that zombie pancreas just wouldn't. The next few days, JB kept pouring out pancreatic juice. I thought about taking him back to the OR, and even talked to him about it.

"I knew you shouldn't have taken my pancreas out," he said.

While I was still agonizing over what to do, I got a text from John, my fellow. "You'll never believe what's coming out of JB's drain now. Succus." JB was leaking from where I'd sewn the bowel back together. And why wouldn't he be? His anastomosis had been bathed in pancreatic juices for three days.

This trip to the OR went only marginally better than the last one.

It seemed as if a bomb had gone off in JB's body since the last time I was in there. I carved my way to the small bowel and was able to take down that anastomosis and put it back together. I then spent at least an hour trying to find and remove any other tissue that looked like pancreas. I called in a partner who had years of experience operating on this beast of an organ, and together we peeled out even more tissue. When I couldn't imagine anything possibly being left, I closed JB up, leaving a drain.

The next morning I looked at the drain—and it looked back at me, laughing. There was the same fluid, in the same quantity. I sent JB down to get a CT scan, and sure enough, I spotted the culprit: a little nubbin of pancreas, smaller than a golf ball, had been left behind. When I showed the scan to my partners, Dr. Sollinger quietly whispered, "El Diablo."

The name had been introduced to Sollinger at a conference he attended in South America, to describe scenarios in which the pancreatic duct gets fully disrupted and disconnects a portion of the pancreas from the rest of the organ, causing a remnant of pancreas to continuously leak pancreatic juices. Sollinger told me I would be back in the OR.

I had decided to wait it out for a few more days when I got another text from John. "Guess what's in the drain now," he said.

I could only imagine.

"Shit," he said. "There's shit in the drain."

Back to the OR, take three. This time El Diablo had spewed its venomous juices through the wall of the colon. We removed a large piece of colon and brought up an ostomy. JB would wear a bag now. We carved out more tissue that looked like pancreas, and I again had a partner come in and help. Sure that it was all gone now, I left a drain. When we went to close, we noted the poor condition of

JB's fascia, the strength layer around the muscles that keeps surgical wounds intact; from all the operations and the leaking stool
and toxic juices, it had been eaten away. He was definitely going to
develop a wound problem.

Next day: bedside of JB, drain, El Diablo. I went back to my office
and put on the Rolling Stones' "Sympathy for the Devil." I listened
to the words, feeling more and more helpless.

> *Pleased to meet you,*
> *Hope you guess my name.*
> *But what's puzzling you*
> *Is the nature of my game.*

How the hell was I going to solve this?

A few days later: back in the OR with JB and more leaking bodily
fluids. This time, I couldn't even get into his belly. Everything was
stuck together. Eventually, we cored a small tunnel into the space
where I thought the remaining pancreas might be. I was nearly
crazed, pulling out tissue to send to Pathology and thinking, *Die,*
you fucking zombie, die! By the third sample, the tissue finally came
back positive for pancreas. We continued to send tissue to Pathology until it came back negative again. It just *had* to be gone now.
And it was.

The next day the drain showed no output, which was good news.
But his kidney had failed, and he was back on dialysis. He had bed-
sores and couldn't walk. Still, slowly, over months, he healed. He
finally made his way to a nursing home. While there, he developed
a nonhealing ulcer on one of his feet, which led to a below-the-
knee amputation. Yet, somehow, he got through all this and made it
home. Although his kidney function was marginal, he was weaned

off dialysis. Six months later, I got word that he wanted to see me in the clinic. When he showed up, he was barely recognizable. He was as thin as a rail, looked about ninety years old (he was sixty), and was missing a leg, but he was smiling. He thanked me for saving his life.

Then he asked me, "When can I get another pancreas?"

That devil had nearly killed him. How could he entertain another? I told him he really wasn't a candidate (he was too old and had been through too much), and no one would want to operate on him.

He told me he wasn't surprised, but then said, "I wish you'd left it in. I don't want to be diabetic again. I would rather have died." And he did, about six months later.

You will never meet a more grateful patient than one who has received a new pancreas and now no longer has type 1 diabetes. No matter what complications that damn devil may cause, patients almost always want another one. Sure, the kidney is the real lifesaver; it gets patients off dialysis, a living hell. And of course, kidney patients are grateful, too. But there is something special about the pancreas. It truly is a bedeviling organ.

The Future of Pancreas Transplant

Outcomes after pancreas transplant have improved dramatically, thanks to better immunosuppression and the commitment and persistence of a handful of investigators. Both the University of Minnesota and my own program at the University of Wisconsin have performed more than a thousand such transplants each. Graft survival at one-year rivals that for kidneys, and we have many patients who are twenty years post-op and doing great.

That said, this is one area where technology undoubtedly will surpass the challenges and benefits of organ transplant. The first-ever closed-loop pump system, which monitors a patient's blood sugar every five minutes and then responds to low levels by lowering the basal rate of insulin, was recently approved. While this device's sensor is subcutaneous (that is, under the skin), I'll bet that in the near future there will be devices that can monitor glucose continuously through transdermal sensors (or some other noninvasive way, such as through contact lenses, a strategy that is being developed). Eventually, these devices will be able to constantly adjust someone's insulin dose to maintain near-perfect glucose levels throughout the day. I presume they will be controlled by smartphones, so perhaps if a patient wants to eat a piece of cake, he can take a picture of the cake with his phone, which will then calculate the needed bolus of insulin and deliver it as the cake is being eaten. Whether that will prevent the long-term complications associated with diabetes will take years to determine.

In the meantime, pancreas transplantation remains one of the best things we do in transplant. No words can adequately express the relief patients get from the burden of disease when they receive a new pancreas. After a life of painstakingly watching their glucose levels, limiting their diets, carrying around insulin and sugary snacks, worrying about passing out or suddenly having a seizure, and waking up every two hours to check their sugar levels, a new pancreas sets them free in a way no other transplanted organ does.

| 9 |

Prometheus Revisited

Liver Transplants and Thomas Starzl

The myth of Prometheus means that all the sorrows of the world have their seat in the liver. But it needs a brave man to face so horrible a truth.

—FRANÇOIS MAURIAC, IN *LE NŒUD DE VIPÈRES* (1932)

Be calm and strong and patient. Meet failure and disappointment with courage. Rise superior to the trials of life, and never give in to hopelessness or despair. In danger, in adversity, cling to your principles and ideals. Aequanimitas!

—WILLIAM OSLER

Madison, Wisconsin, 2:30 a.m.

"Careful, Bobby. Those veins will bleed like stink."

I craned my neck to watch my fellow Bobby carefully dissect the liver off the vena cava. I couldn't really help him because I was using both my hands to hold up a huge cirrhotic liver so he could work under it. My back was killing me, and my left arm was going numb. We had been in the operating room for about three hours at this point, and still had a long way to go. We'd already dissected out the upper cuff, divided the vessels and the bile duct that go into the liver, and were now peeling the liver off the cava.

The vena cava is a gigantic vein that carries blood from the whole body back to the heart. The liver hugs the cava tightly, with engorged blood vessels that drain directly from the liver to the cava. Sometimes there can be fifty veins going into the liver, and tearing any of them can lead to torrential bleeding. Of course, this is something we get used to in this business, which is why we have two suction devices going at all times.

Bobby, who was toward the end of his fellowship, did a masterful job peeling the liver off. At this point, we were almost ready to take it out; the only structures holding it in were the three large hepatic veins that form the upper cuff. This liver had a TIPS catheter in it, shorthand for a transjugular intrahepatic portosystemic shunt, a giant straw that radiologists had snaked into it, through the right hepatic vein, to reduce the resistance to blood flow in this hard, cirrhotic liver. By allowing the blood to flow through the liver, the TIPS dried up all the fluid that had accumulated in the patient's belly while he was waiting for a donor organ.

I managed to get clamps around the cava, although I could tell my upper clamp was partially on the TIPS, preventing it from being pulled out. I made a cut into the right hepatic vein, and there was the TIPS, staring at us. I got it better exposed, and Bobby got a clamp around it.

"You ready, Bobby?"

We had rehearsed our moves prior to the case, and he tightened his grip on the clamp. I asked anesthesia to lower the patient's head, to prevent air from entering the pulmonary artery or traveling to the brain if there was an air embolism. Although rare, this deadly complication can occur when you open a large vein, like the hepatic vein, if the pressure of air outside the vein is higher than the pressure of blood returning to the heart, which can be the case in liver surgery. Lowering the head can allow gas bubbles to rise up in the heart rather than travel out the pulmonary artery.

"Here we go," I said calmly, and opened my upper clamp. As blood started to well into the field, Bobby pulled hard on the clamp. The TIPS rose out of the heart, and I reclamped the cava. Before I could celebrate, though, I heard someone from anesthesia say over the drape, "Uh-oh, big clot in the heart"—they could see it through their ultrasound probe—"He's arresting. Better start CPR."

Damn. Maybe I was being too much of a cowboy on this one. We pulled the retractors down, and Bobby started CPR. While this was going on, all I could think about was having to talk to the patient's wife and kids. I could picture their faces the moment I told them he hadn't made it.

Despite that vision, I felt weirdly calm and disconnected. After about ten minutes, with all of us covered in sweat, the patient's heart kicked back in. His blood pressure returned.

We stood there for a minute, shaking, not sure what to do. *Do we get the donor liver out and sew it in? Does his brain still work? Or would that be a waste of an organ we could give to someone else?*

Bobby broke the silence. "Open the new liver. Give me a stitch."

"Okay, let's do it," I said.

Don't worry. Everything turned out fine. The patient's a veteran. You can't kill a vet.

Pittsburgh, 2016

My alarm was set for 6:00 a.m., but I was up by 5:00. It wasn't my insomnia this time; I was just excited. It's not every day you get to meet one of your heroes. I got out of bed, opened my backpack, pulled out my well-worn copy of *The Puzzle People*, and turned to the first chapter. While I had read the book many times, I wanted to make sure I had the early years down pat. I figured the more I knew about Dr. Thomas Starzl, the more likely he was to open up to me. I had corresponded with him numerous times this past year, but it had taken quite a while for him to agree to me coming to visit. He was more than willing to send me copies of articles he had written about transplant, and summaries already available in the medical literature about the founding fathers of the field, but I wanted something different from him. I wanted to understand how he had been able to do what he did: to persist in a field that no one thought possible, where his initial patients, many of them children, were all dying on the table, and to drag the field kicking and screaming into a clinical reality. He did all this despite much resistance and with fellow clinicians signing petitions for his removal and calling him a murderer. Sure, it is easy to say that those patients would have died anyway, so it's not that big a deal. But I don't buy that for a second. Having had a patient die while in my hands; having had to walk out of the OR while everyone was standing in silence looking at me and the patient was lying there cold and lifeless, blood pouring out of the incision and onto the floor; having been the one to have to tell the family about their loved one's death—I don't see how it helps to know that the patient would have died anyway.

Was Tom Starzl made of the same flesh and blood I was? For the longest time, I assumed he was not. Then I found, hidden away in his memoir, tucked inconspicuously on pages 59–60, a surprising and fascinating quote:

The truth was worse than anyone imagined. For the past six years, I had honed my surgical abilities. At the same time, I harbored anxieties which I was unable to discuss openly until more than three decades later, after I had stopped operating. I had an intense fear of failing the patients who had placed their health or life in my hands . . . Even for simple operations, I would review books to be sure that no mistakes would be made . . . Then sick with apprehension, I would go to the operating room, almost unable to function until the case began.

He went on to write:

Later in life, when I told close friends that I did not like to operate, they did not believe me or thought I was joking. Most surgeons whom I know have been able to protect themselves, either by rationalizing errors which they had committed or by promptly erasing the bad memories. I could not do this. Instead of blotting out the failures, I remembered these forever. With growing concern, I came to believe that I was not emotionally equipped to be a surgeon or to deal with its brutality.

How could someone who felt this way actually have selected the path Starzl walked? Why did he choose to master an operation that no one else could do, one that would surely lead to the death of everyone he operated on? And not just death, but a bloody morass like the aftermath of a crime Jack the Ripper would have admired?

I needed to try to understand this better. But would he let me in? Also, Starzl had recently turned ninety. Would he remember those early days of transplantation? Would he have access to all the emotions he felt when starting out?

Later that day, I stood outside a building across the street from a huge construction site looking at a rather dingy doorway nestled between the Campus Bookstore and the Prince of India on Fifth Avenue in Pittsburgh. I don't know what I'd been picturing. Maybe a glistening gold door with a carved handle? At the very least, glass doors in the wing of the Transplant Institute. Hell, the entire Transplant Institute at the University of Pittsburgh had been named after this guy, and here he was, tucked away in this dilapidated building. Well, such is life in academic medicine.

I was buzzed in and began climbing the steep stairs to the second floor. Standing at the top was a smiling elderly man in a blazer and tie. A few steps up, I stepped in a moist puddle of vomit, and Starzl said, "Is that vomit down there? Something must be going on with Chooloo."

As I got to the top of the stairs, I saw the perpetrator, a golden retriever looking over to see who was coming in. I remembered that Starzl was a dog lover, even though he had sacrificed so many dogs in his early years of transplant trying to perfect the operation.

He must have guessed what I was thinking. "The one thing I know, I love dogs," he said.

We went into his office, a large room with creaky floorboards and sparse furnishings: a few folding chairs, a large table pushed up against a wall and covered with manuscripts and papers. But it was on the well-stained couch in the corner where this giant of transplant began his story.

Thomas Starzl was born on March 11, 1926, in Le Mars, Iowa.

His father owned and ran the local newspaper and was also a fairly successful and prolific writer of science fiction. Dr. Starzl's mother had worked as a surgical nurse before he was born, which clearly played some role in his decision to become a surgeon. He grew up under the shadow of the Great Depression, and developed a belief in hard work, an ability to put his head down and work for hours on end, and a sense of responsibility to his family and the people around him. Starzl was not a man who would complain about his lot in life, but he desperately wanted to leave small-town Iowa.

After a stint in the navy, where an aptitude test suggested he would thrive as a physician, Dr. Starzl entered Northwestern University Medical School, in Chicago, in 1947, and then spent four years at the Johns Hopkins Hospital. He found the experience there brutal. Back then, surgical programs were pyramidal, meaning a majority of residents would ultimately be fired. After four years, Starzl was notified he would not be able to complete his training at Hopkins—which was okay with him. He was ready to get out of Baltimore.

He spent the next two years at Jackson Memorial Hospital, at the University of Miami, where he performed roughly two thousand surgeries. Somehow, he still managed to set up his own lab, and it was here that he began his first research project examining the liver.

The liver is a unique organ in that it has the ability to repair itself after injury. Unlike the cells of any other solid organ in the body, hepatocytes, or liver cells, can grow in size or divide. If you cut out half of someone's liver, the remaining half will regenerate within weeks to months. *Regenerate* may not be the right word; it doesn't actually grow back as you might imagine a tail doing on a lizard or a frog. But through a combination of cell growth (hypertrophy) and division, the liver will regain its original size and function. This phenomenon is truly amazing.

Nonetheless, if enough injury occurs over enough time, the liver can shrink and be totally replaced with scar tissue, losing its regenerative ability as its architecture gets more and more disrupted. Once it reaches this end-stage state, it is termed *cirrhotic*. The problem with cirrhosis is twofold: First, the liver can become dysfunctional, no longer able to make the various proteins and clotting factors it is charged with making, and no longer able to detoxify the various wastes or break down the products that pass through it on their way from the intestines. This can cause confusion in patients and spontaneous bleeding. The liver can also lose its ability to make and drain bile, so patients turn yellow (become jaundiced). Second, a knotty, shrunken liver (which is what most cirrhotic livers become) can impede blood flow. Despite its normally high flow, in cases of cirrhosis, the portal vein backs up and can't push blood through this highly resistant organ. This backup of blood leads to the distension of this already huge vein and the reversal of flow away from the liver. When this happens, many small veins coming off the portal system get distended, becoming varices (literally, dilated veins). This includes the veins that run along and into the esophagus; these veins can spontaneously rupture, an extremely life-threatening condition and one that can be very dramatic: patients can show up at the hospital vomiting massive quantities of blood and die right in front of you. A second complication of the backup of blood flow in the portal vein is ascites, or (nonblood) fluid that accumulates in the patient's belly, sometimes ten liters or more. This leads to the characteristic appearance of a patient in liver failure: yellow, swollen, and often looking nine months pregnant.

There weren't many options for patients presenting with cirrhosis and esophageal bleeding in 1955. A number of operations had been designed to divert blood around the liver as it took its path from the organs of the gut back to the heart. Starzl was involved in one of these operations, in Miami, and he was surprised to note that when the patient's liver was bypassed, his diabetes was cured. Starzl was fascinated by this case and decided he would study it in animals. Disappointed to find that there was no large animal facility at Miami, he set one up in an empty garage located near the hospital, "borrowing" equipment from the hospital to make it work and enlisting his wife and a junior resident to care for the animals. They got dogs from the city pound, made them diabetic using chemicals toxic to the cells in the pancreas that make insulin (beta cells), and then performed these same operations, bypassing blood around the liver. Lo and behold, the diabetes . . . got worse.

Starzl felt that his hypothesis could not really be tested without removing the liver entirely, so he developed a technique for total hepatectomy. Of course, the dogs could survive only a day at most, but this was Starzl's first step on the path to liver transplantation. "The most important consequence of the liver removal operation was the realization that a new liver could be installed (I thought quite easily) in the empty space from which the normal liver had been taken out," he would later write in *The Puzzle People*. "In fact, half the operation of liver transplantation already had been perfected with the hepatectomy procedure. The other half would be to sew in a new liver." At this point, he was hooked.

AFTER HIS TIME in Miami, Starzl returned to Northwestern in 1958. He decided to take an extra year of training as a fellow in chest surgery. He knew his passion was the liver, but despite being gifted at surgery, he was tortured by the idea of performing it, and much preferred research. During his fellowship at Northwestern,

he conducted liver transplants in dogs in his lab (with no assistants). All the dogs died. This would be harder than he'd expected.
As he completed his last year of training, he agreed to stay on at
Northwestern to continue his research. After obtaining a couple of
national grants, he was off. Not only did he develop the techniques
in dogs that would be necessary for attempting liver transplantation
in humans, but he also was finally exposed to the handful of other
investigators working in the field of liver transplantation.

At the Brigham more or less at the same time, Francis Moore also
had turned his attention to the liver. He put a group together and
performed numerous liver transplants in dogs. Early attempts failed
for him, too, as the dogs did not tolerate the clamping of the vena
cava and portal vein that was required at the time to get the liver out.

Both Starzl and Moore came up with the same solution to this
problem. They obtained plastic tubing and shunted blood from the
lower extremities into a vein in the neck that drains into the heart.
In this way, when they clamped the cava, blood would bypass the
clamped vein and make it back to the heart. (Their technique is
still used today in some programs.) Once this problem was solved,
both groups were able to transplant the liver in dogs successfully
and saw slowly improving survival rates. Of course, neither group
was using immunosuppression, as there was none at the time. But
even though all the livers were rejected in about a week, during that
week, the labs normalized, suggesting the livers were functioning,
and the dogs behaved normally.

At the annual meeting of the American Surgical Association,
Franny Moore presented the Brigham data. His description of liver
transplantation was seen as groundbreaking—until Tom Starzl got
up to discuss Moore's paper. As Starzl recalls:

My approach had been and would continue to be strongly influenced by my original interest in the effect of portal blood (and insulin) on the liver. In more than eighty transplant experiments, I had systematically tested different ways of restoring the transplanted liver's blood supply. We showed that livers which were given a normal portal venous inflow performed better than those which were not. But the important achievement for the moment was that we had eighteen dogs with survival greater than four days, with one animal living for twenty and one-half days. I realized that we were ahead of the Boston team.

It was at these national meetings that ideas were shared and, slowly but surely, Thomas Starzl gained recognition as the central force in efforts at liver transplantation. Indeed, it was at the ASA meeting that Starzl first heard about ongoing experiments (both at Richmond with David Hume and at the Brigham) using chemical immunosuppression with a new drug called 6-Mercaptopurine. It became clear to him that chemical immunosuppression was the next step in transplant, and that in order to make liver transplant a reality in humans, he would first need to master kidney transplant and move forward with chemical immunosuppression.

In December 1961, Starzl moved out to Denver, where he was appointed chief of surgery at the Denver Veterans Affairs Medical Center. Within three months of his arrival, he performed his first kidney transplant. Given the generally bad outcomes that were the norm during this period, he was smart enough to start with a pair of identical twins: he felt he had to prove his worth to the local physicians and medical community with at least one good outcome before diving into the insane world of immunosuppression in the 1960s.

By the time Starzl got into the kidney business in 1962, Joseph

Murray at the Brigham had already reported his successful case with nonidentical twins and had just introduced the immunosuppressant azathioprine to human transplantation. It would be this same year that he would have his first graft survival with a deceased-donor transplant.

Although Murray was proceeding as quickly as he could, and Hume and Roy Calne were performing transplants at their centers, Starzl jumped in like a freight train. Calne and Murray had identified that azathioprine would be useful in kidney transplantation, but it was Starzl who realized that combining it with high doses of steroids was a better strategy than using azathioprine alone. Many investigators contributed during these early years of transplant, but Starzl, with his large surgical volumes, his ability to push new protocols through uninhibited, and his obsession with writing up his results, played an outsize role.

After his successes in the kidney world, Starzl knew it was time to try his hand at clinical liver transplantation.

Prometheus Rising

I love that feeling when you pop through the peritoneum, encounter liters of beer-colored ascites, and first lay eyes on a shrunken liver. You never know what a transplant is going to be like until you get in there, but the second you see the shrunken, mobile liver floating in ascites, you know you're in for a treat.

"Okay, guys, crank the Pitbull. If I don't screw this up, we should be done in four hours, maximum."

Liver transplants used to be like this ten years ago, back when it was so much easier for our patients to get one. Nowadays, thanks to changes in allocation that have increased the sharing of livers around the region, our patients have to be so much sicker to get these precious organs. I suppose that's fair, but it sure has made my job harder.

"C'mon, Elliot," I said—Elliot was in the second year of his fellowship, having already completed his surgery residency—"let's move this along. I really want to get out of here today."

We had the liver out in just over an hour. "Open the new liver," Elliot said as we switched sides. Elliot is a lefty, so we had to switch in order for him to sew the upper caval anastomosis.

We pulled up the new liver, glistening and perfect—such a contrast from the knobby, shrunken liver we'd just passed off the table to be sent to Pathology. I held it out of the way while Elliot began to sew in the upper cuff. At this point we didn't really need to say anything about the case. Elliot knew every step and exactly how I wanted it done. This one went perfectly. He sewed the back wall first, and then the front, throwing each stitch while I "followed" him, holding on to the suture to make sure it didn't get tangled. Once he reached the end, he tied the two ends up, and we moved on to the portal vein. Again, I assisted him as he sewed that end to end with three 6–0 Prolene sutures. It took us about thirty-five minutes to sew this new liver in, and then we removed our clamps and reperfused. The liver pinked up beautifully, and the patient tolerated it without turning a hair. We then turned our attention to the hepatic artery, which again lined up nicely and seemed almost too easy. I looked up at the clock and smiled. It was only 6:30 p.m. I should be able to make it home before my kids went to bed (my daily goal).

"What's so hard about liver transplant?" I said to Elliot. "I wonder why Starzl had so much trouble?" We both laughed, thinking about the last one we did, which took twelve hours. You just never know.

ON MARCH 1, 1963, Thomas Starzl attempted to perform the first liver transplant in a human. The recipient was Bennie Solis, a three-year-old boy who was unlucky enough to have been born with a disease called biliary atresia, in which the ducts that normally coalesce to form the common bile duct, allowing drainage of bile out of the liver and into the intestine, never form. Perhaps the worst part of this disease for the parents is that those beautiful babies suffering from biliary atresia appear normal when they are born—with the exception of neonatal jaundice either at birth or developing after a couple of weeks. Since half of all children develop neonatal jaundice that eventually resolves, Bennie's parents were likely told not to worry. Eventually it must have become obvious that something was wrong—maybe he had some white stool or dark urine; maybe he became inconsolable because of persistent itching; or maybe the yellow color just wouldn't go away. Lab tests would confirm everyone's worst fears.

Back then, a diagnosis of biliary atresia was essentially a death sentence. By the time Starzl met young Bennie, he was three years old, yellow, swollen with ascites, and filled with thin-walled nests of varices, resulting from the inability of portal blood to flow through the cirrhotic liver. As if that weren't enough, Bennie's liver dysfunction was so severe that his blood contained none of the clotting factors we normally rely on to stop bleeding during surgery.

At the time he met Bennie, Starzl had performed two hundred liver transplants in dogs using prednisone and azathioprine, with reasonable short-term survival rates. He had also performed four nonidentical kidney transplants in humans while using this same immunosuppression, all of which were still functioning for at least four months by 1963. Bennie seemed to be an appropriate liver transplant candidate. In Starzl's own words:

[W]e viewed the principal hurdle to be the operation itself, which would be vastly more difficult than kidney transplantation. However, nothing we had done in advance could have prepared us for the enormity of the task. Several hours were required just to make the incision and enter the abdomen. Every piece of tissue that was cut contained the small veins under high pressure that had resulted from obstruction of the portal vein by the diseased liver. Inside the abdomen, Bennie's liver was encased in scar tissue left over from operations performed shortly after his birth. His intestine and stomach were stuck to the liver in this mass of bloody scar. To make things worse, Bennie's blood would not clot. Several of the chemical and other factors necessary for this process were barely detectable. He bled to death as we worked desperately to stop the hemorrhage. The operation could not be completed. Bennie was only three years old and had not enjoyed a trouble-free day in his life. Now, his wound was closed and he was wrapped in a plain white sheet after being washed off by a weeping nurse. They took him away from this place of sanitized hope to the cold and unhygienic morgue, where an autopsy did not add to our understanding of our failure. The surgeons stayed in the operating room for a long time after, sitting on the low stools around the periphery, looking at the ground and saying nothing. The orderlies came and began to mop the floor. It was necessary to prepare for the next case.

"THIS IS NOT good. This is definitely not good." I uttered this quietly, but I'm pretty sure everyone in the room heard me. When things get really bad in the OR, I tend to speak more quietly. I never yell, but everyone in the room knows things are serious when the jokes stop. They turn the loud music down and try to listen to what I'm mumbling.

It was about two in the morning, and Paul (a second-year transplant fellow) and I had just reperfused the liver. And that's when the shit truly hit the fan.

We had found out about the liver earlier in the day. It was from an older donor, a man in his early seventies whose liver had been biopsied and looked fine. Older livers can work well, but the transplant operation needs to go smoothly—old livers don't like to sit out in the cold a long time and would like their blood flow back as soon as possible.

The recipient, Tito, had been admitted to the unit the night before, and I thought I would swing by and check him out prior to accepting the liver for him. He was sitting in a chair, an oxygen mask on and a catheter coming out of his bladder. His daughter was giving him an extremely worried and caring look. Tito appeared feeble, fragile, exhausted, but he wasn't on a respirator, he was sitting up on his own, and he was able to smile when Paul and I walked in. After a few pleasantries, I told him we had just accepted a liver for him.

The room erupted in cheers. Everyone knew this meant life for Tito, and not a moment too soon. I mentioned that it was a big surgery, that there was a lot of risk, that he could die—the talk I'd given so many times before.

THE HEPATECTOMY WENT pretty well. The portal dissection went smoothly—we divided the artery, portal vein, and bile duct without too much trouble. Then we freed the liver off the cava, gaining control around the cava up high above the liver near the diaphragm and down low below the liver near the kidneys. We were ready to cut it out.

I noticed that the bowels had become considerably swollen, because their venous drainage, which goes through the portal vein, had been clamped off. I knew that might make it harder to sew the new liver in, since there wouldn't be a lot of space to work in, but nothing I could do about it now. I put the clamps on the cava and we proceeded to cut the liver out. Then we opened the new liver (meaning our circulating nurse removed it from the cooler, opened the outer bag, and let our scrub tech pull the inner sterile bag out on the field; we package every organ in three sterile bags for protection).

It was a lot bigger than I was expecting. We pulled it up to the field and started to sew it in. I was concerned with how little room we had to work in, what with the big liver, the swollen bowels, and the diaphragm bowing into our field from the fluid our patient had accumulated in his right chest (a common occurrence in cirrhosis). *Damn, I thought this was going to be easier.* I wondered if we should stop and make our incision bigger, or just go for it. While I held the liver down as hard as I could, Paul sewed the donor and recipient cavas together. Once we finished the anastomosis, I placed a clamp on the liver side of the donor cava and released the cava from my clamp. There was a bit of bleeding, which we easily controlled, but it seemed okay. The patient was more stable now, with his cava unclamped. We turned the music back up, sewed the portal end to end between the donor and recipient vein, and got ready for reperfusion. The anesthesia and nursing teams were ready. We flushed the liver out with saline and then blood, and then released the clamps.

Everything was okay, and then—he started exsanguinating. It was massive. Paul and I put our suckers by the caval anastomosis, but the bleeding was so intense that we couldn't see anything. Hence my quiet statement "This is not good. This is definitely not good."

I quietly told anesthesia we were in serious trouble, and asked the nurses to call in one of my partners. Somehow, we were able to get our clamps back on the cava and the portal vein. Now the donated liver was getting no blood flow.

Once everything was clamped, and the bleeding had stopped, what we saw was . . . well, not good. The entire upper cuff of the cava of the donor liver had become shredded. There were multiple linear tears (meaning the sutures had pulled through all the way around, causing long straight tears in the cava), with more air between the sutures than tissue. We were fucked. I probably needed to take the liver out and try to fix it on the back table, perhaps by getting some vessels from the vessel bank and then trying to sew the liver back in. But as I was considering this, Sergei, the anesthesiologist, told me that we were in serious shit. Tito, our patient, was going to code at any minute. I started to picture Tito's daughter. She had been so happy when I told her we had a liver for her father.

By now my partner Dave had joined me, appropriately impressed with the situation. We racked our brains over what we could do, other than just stand there and watch Tito die. And somehow, we came up with an idea, one that seemed so crazy it just might work, although we had never done it before.

I placed a side-biter clamp on Tito's cava down below the liver. I cut a hole in his cava above my clamp and proceeded to sew the donor infrahepatic cava to the recipient cava. This took about ten minutes. We opened the clamps and there was flow through it. Then I grabbed a vascular stapler and fired it across the upper cuff. Most of the surgical bleeding from the cava stopped. Success. We basically rerouted blood, so that rather than flow through the donor liver and out the top, the blood flowed through the liver and went out the bottom, still into the recipient cava.

Except . . . we were still swimming in blood. It was now about five in the morning. We had been working all night, and the patient had been tanking for two hours.

"This is not working," Sergei said, stating the obvious.

The donor liver looked like dogshit—it was mottled, swollen, and pale, and Tito's labs were abysmal, making it unlikely that he would survive no matter what we did. This new liver was not functioning at all. I told the nurses to find out where his family was; I wanted to tell them he wasn't going to make it out of the OR. Somehow it seemed important to tell them this before it was over. At least he was still alive now.

It is a helpless feeling walking out of an operating room knowing your patient is going to die. I couldn't stop thinking that maybe if someone else had performed the surgery—maybe Tony, or Dave—this wouldn't have happened. I walked toward the surgical waiting room; I could see Tito's daughter, Orinda, in the distance. I was so tired I could barely walk, but I was acutely aware that she was staring intently at my face, trying hard to catch some clue as to what I was about to say.

I sat down next to her. "Things are not going well. We initially got the liver in, but then we had a tear in the blood vessels. Your father has lost a lot of blood, and he is very unstable. I really don't think he is going to make it out of the operating room."

There it was. I'd gotten it out. I could see tears in her eyes, but she held it together.

"Is he still alive?"

"Yes," I told her. "But he is very sick. I don't know if his brain is okay. The liver is not working. I think you should call your family to come in now."

She thanked me profusely for having done everything we could, but then she said, "I know you will do everything you can to try and save my dad."

Those words were ringing in my ears when I got back to the OR. Maybe we could at least make some progress, I figured; get Tito to the ICU, get him to a point where we could at least entertain the idea of a new liver.

I asked for a stitch, and aggressively threw it in a small hole in the cava. I asked for another. And another. *Fuck it. Let's do this.*

Dave, Paul, and I spent the next three hours throwing stitches, burning tissue into bloodless submission with the heat of the argon beam, and intermittently packing Tito's belly. The energy in the room started to change. We started to think that maybe there was a chance. It still seemed like a one-in-a-million chance, but it was a chance. Tito had lost close to a hundred liters of blood, which, while not a record, is astronomical. There was still a lot of work to do, but I told the nurses to get Tito's family into a meeting room. I wanted to give them another update.

"Okay, here is the deal," I told them when I found them later. "We have definitely made some progress. But Tito is very sick. I still think it is likely he won't make it through this. I have no idea if his brain is okay. We won't be able to close his belly for now, and he will definitely need to come back to the OR. Honestly, the absolute best-case scenario is that we make it to a point where we can list him urgently for another liver. But even that is a long shot."

At this point, there were about twenty family members there with Orinda. They seemed understanding, and said they were very thankful. They said Tito was a fighter. I felt so glad that at least they'd be able to say good-bye to him.

I went back to the OR. By this point, Tony and Luis had scrubbed in, replacing Dave and joining Paul, who was now only barely awake. (He was in for the long haul. Such is the life of a fellow.) Tito was in good hands.

I walked to the locker room, sat on a bench and looked down at my scrubs. They were covered in blood. I peeled them off and threw them in a hamper. I barely remember driving home. Once there, I stumbled up to my bedroom. My dog, Phoebe, seemed confused. She ran behind me, no doubt hoping I would take her out. I envied her. She has a pretty good life: she sleeps all she wants, and she doesn't kill anyone . . . except maybe a squirrel or a rabbit.

I lay in bed, my head spinning. I could still see all that blood welling up in Tito's belly as I fell into a dead sleep. Two hours later, there was a text on my phone telling me I had a patient to see in the clinic.

I SAW TITO in my office recently with his daughter, Orinda. After the transplant, he was extremely ill in the ICU. His kidneys failed, he was on a ventilator, and the liver we had put in was barely keeping him alive. After a week or so, we got him a new liver. That transplant went better. He spent about a week in the ICU and another few weeks in the hospital. Then he went to rehab, and now he was home. His new liver was perfect; his kidneys recovered, and he was back with his family, looking great. (Thinking back on the operation, I couldn't believe this was the same man sitting in front of me.) He told me about his childhood in Puerto Rico, his current life in Wisconsin. He told me about his big family, how much he loved them. Orinda was sitting next to him beaming, his guardian angel.

The thing that blows me away about Tito's case is how close I came to giving up on him. I remembered the thoughts that went through my head in the OR:

I can't save him.

He doesn't have a chance.

It would be better for his family if we called it and moved on.

They could grieve and then go on with their lives.

But then something changed my mind. I attribute it mostly to Orinda's expression of her trust in me. This is the crux of being a surgeon: Patients and their families trust you to do the best thing for them, for their loved ones. They take you at your word. You bring people to the OR, you open up their bellies or their chests, their heads, their limbs, you put their lives into your hands, and your judgment becomes everything. This is both the beauty and the challenge of surgery. This is why we train so hard and help one another and push ourselves to be perfect even when we can't be. This is why surgery can be so wonderful, but also so humbling. It is also why we need to call for help, own our mistakes, always try to be better.

These organs we transplant—the livers, the kidneys, the hearts— they are the ultimate gift, the gift of life, the last thing the dead can bestow upon the living. We, as surgeons, simply transfer them from one person to another. We are the stewards, and it is our job to make sure the gift is given. Yet I am blown away by how hard we need to fight for it sometimes, how often we must remind ourselves about the pioneers in transplant: They never gave up. It would never have crossed their minds.

THOMAS STARZL WAS not deterred by the surgical challenge presented by the case of Bennie Solis, the three-year-old who died on the table. He recruited a clotting expert onto his team and came up with a game plan that involved infusing human clotting factors during surgery. With this technique, Starzl performed his next transplant in May 1963, on a forty-eight-year-old male with a primary liver cancer. The surgery was successful, and the patient awoke by the next day with a functioning liver and beautiful golden bile coming out of the tube left in his bile duct during the surgery. Unfortunately, twenty-two days later he died when blood clots, which had formed in the plastic tubing bypassing blood around the liver while the portal vein was clamped, entered his lungs. Starzl would perform three more technically successful liver transplants in 1963, but all these patients would die from clotting problems.

In September 1963, Franny Moore and his team jumped into the fray, performing the first liver transplant at the Brigham, on a fifty-eight-year-old patient with metastatic colon cancer. While the patient made it through the operation, he died on post-op day eleven from pneumonia and infection in the liver. An attempt was also made in Paris, in January 1964, but this failed due to hemorrhage. Moore and his team completed four cases, two adults and two children, between 1963 and 1965. None survived. Soon clinical trials were halted all over the world due to a self-imposed moratorium on the procedure.

Starzl was certainly disappointed with these early failures, but in no way did they shake his belief in liver transplant becoming a reality. He was a man on a mission, and he was going to take this all the way. He knew there would be heartbreak, complications, and resistance from colleagues. None of that deterred him. By this point he had done hundreds, maybe thousands, of transplants in dogs and knew the operation on a healthy animal inside and out. He also knew that things such as coagulopathy and infection would need to be sorted out in humans as well. His lab work became very focused.

Three months before he resumed liver transplantation, he decided to focus only on children. The pediatric department at Denver supported his efforts, but the adult medical team did not. This was not the first time Starzl proceeded without support, and it wouldn't be the last.

On July 23, 1967, he transplanted Julie Rodriguez, a nineteen-month-old with an unresectable cancer in her liver. The surgery went beautifully, and Starzl thought he had cured her—until new spots were noticed on her chest X-ray three months later. He returned Julie to the OR at three and a half months and again at seven months, to take out new tumors, but they kept recurring. She lived for four hundred days. Starzl would keep a painting of her over his bed for the rest of his life.

On April 17, 1968, Starzl presented data on his first seven transplants since the moratorium had ended. Four patients had died in less than six months, but three were still living. One was Julie; a second was Terry, a sixteen-year-old who lived for more than a year and actually returned to high school until her tumor recurred. The last, Randy, was a two-year-old with biliary atresia. After suffering chronic rejection, he died at the age of four and a half during an attempt at retransplantation. In 1969, Starzl finally had a long-term success. Kimberly, another child with biliary atresia, was still alive twenty-two years later.

Starzl put it bluntly: "A grim conclusion was unavoidable. Liver transplantation was a feasible but impractical way to treat end-stage liver disease." The bad outcomes continued to weigh heavily on him:

The mortality from the failed early trials and that which occurred later did not mean that liver transplantation was causing deaths. These patients were under a death sentence already because of the

diseases that had brought them to us. Even now, I continue to re-
ceive letters from parents or family members. These always start by
saying that they know I won't remember Jimmie or whatever was
the patient's name. Then they express thanks for the fact that we
had made an effort instead of letting their children die, off in a back
room without hope. Those opposed to trying always claimed that
these little creatures had been denied the dignity of dying. Their par-
ents believed that they had been given the glory of striving.

They were wrong about one thing. That I would not remember.

He remembered every single one of them.

Enter Roy Calne, Again

If Starzl came to the discipline of liver transplantation because he
was fascinated by the organ and the complexity of the operation
(and was looking for a mission), Calne came to it because of his
love of immunology. He began experimenting with liver transplantation in pigs and was surprised to see that rejection was much less
likely with this organ than with kidneys, and in some animals, he
could even achieve long-term graft survival. Finally, in 1968, he
announced to his colleagues at Addenbrooke's Hospital, in Cambridge, England, that he was ready to perform a liver transplant in a
human. His first case would be a woman with a tumor in her liver.

On May 2, a child was admitted with viral meningitis and was
deemed to have an irreversible brain injury. The ventilator was to be
turned off, and the family consented to organ donation. Calne knew
that he was alone in his desire to proceed, and called a meeting with
his colleagues to try to win them over. In yet another stroke of

luck, on top of having a donor and a recipient available at the same time, he received a call from none other than Franny Moore, his old mentor. He was in town visiting his son, a graduate student at Cambridge, and agreed to come to the meeting to discuss whether Calne could proceed with the surgery.

The meeting convened a few hours later. Calne presented his plan, and then went around the room asking for opinions. Every last person voted no; they deemed the surgery too dangerous. Then Calne turned to the visitor sitting quietly at the back of the room, unnoticed but listening intently. "After listening to this litany of pessimism, I introduced Dr. Moore, world-famous (even in Cambridge), and together with Starzl, one of the pioneers of liver grafting. Moore's response was short and typical of him. 'Roy, you have to do it.' The opposition collapsed, and we made immediate plans for the surgery."

Moore assisted Calne on this first liver transplant in England. The case was complex, but ultimately went well. Calne was concerned that the small size of the cava would be too small to replace the recipient's adult-size cava, so he performed the world's first piggyback transplant, leaving the recipient IVC intact and sewing the end of the donor IVC to the side of the recipient IVC. The recipient woke up postoperatively with immediate liver function. Sadly, she died of infection over two months later.

Four of Calne's first five patients never left the hospital. But one, a forty-eight-year-old woman named Winnie Smith, survived for five years (before succumbing to infection after a blockage of her bile duct). She did so well that, in 1972, Calne brought her with him to the International Transplant Society meeting in San Francisco. After her case was presented, she took the stage and answered questions.

In a recurring theme, if the 1960s were full of promise, accomplishment, and hope, the '70s were generally a period of carnage interrupted by the occasional bright moment. Although a few other transplant teams popped up and then folded over this decade, a large majority of the liver transplants were performed by Starzl's and Calne's teams, with Starzl's bearing the biggest load. By 1975, few other liver transplant programs existed. Starzl and Calne, along with kidney transplant surgeons scattered around the world, were using the triple drug therapy of azathioprine, steroids, and an antibody to immune cells that had recently been developed (antithymocyte globulin, or ATG). Still, they lost their patients due to either rejection or, more often, infection. The spark that was needed to truly start the fire was a novel immunosuppressant.

When Calne finally introduced cyclosporine to kidney transplant, Starzl jumped on board to see if this was the spark they'd been looking for. As soon as he had some experience with cyclosporine in kidney transplants, he quickly moved on to liver transplant. This was no easy task. By this point, he was on his way out at Denver—when it came to Starzl and his insane quest to make liver transplantation work, the leadership and most of the faculty there had lost their nerve and weren't interested in supporting another expensive trial that was sure to fail. Starzl hammered through anyway, beginning a liver trial in March 1980. He performed twelve liver transplants between March and September 1980. Eleven of the twelve lived for a year or longer.

In 1981, Starzl published his one-year follow-up on the cyclosporine livers in *The New England Journal of Medicine*. He was at Pittsburgh now, finally starting to believe that liver transplantation might be more than an experimental therapy. But his first four liver transplant patients at Pittsburgh died within four to twenty-two days. Then Starzl's luck—and that of his patients—turned around. Of the next twenty-two liver transplants, nineteen had long-term survival, and that number would keep doubling. By 1983, cyclosporine would finally get FDA approval, and the world of transplantation would explode. That same year, with some help from the US surgeon general C. Everett Koop, liver transplantation was deemed a service rather than an experimental therapy, which led to insurance companies finally covering those undergoing the procedure.

If this were a movie, Starzl would now walk off into the sunset as the credits rolled. But he wasn't done yet. He wanted to train the next generation of transplant surgeons, and he had plenty of organs to work with. As there were still few transplant centers to compete with, the volume at Starzl's hospital skyrocketed—one year, he transplanted more than five hundred livers—and fellows flocked to Pittsburgh to spend two years with the master. It's shocking to think about all the people Starzl trained. Virtually every liver transplant center in this country can trace its origins to Starzl within one or two generations, and to this day, many of the leaders in our field are Starzl disciples. He is like a god to so many, the most important mentor in their professional lives—although more than a few have some PTSD from the experience, the demands and intensity were so high.

Starzl hung up his scalpel in 1991, at the age of sixty-five. Some of his colleagues and trainees were shocked and disappointed, but for Starzl, it was perhaps the easiest decision he had made in years. "I was wiped out," he said. He had accomplished everything he had set out to do, including championing a new immunosuppressive drug, FK506, which would go on to replace cyclosporine as the primary immunosuppressant and is still used today in solid organ transplantation. Starzl lived an additional twenty-five years, and for that time, he came into work every day to continue his research on transplant immunology. He died on March 4, 2017, just a week before his ninety-first birthday. As Münci Kalayoğlu, a Turkish-born surgeon who trained with the "maestro," once said, "Watching Dr. Starzl was . . . was not surgery, but art."

Part IV

| The Patients |

If the history of medicine is told through the stories of doctors, it is because their contributions stand in place of the more substantive heroism of their patients.

—SIDDHARTHA MUKHERJEE,
THE EMPEROR OF ALL MALADIES

| 10 |

Jason

The Secret Is to Live in the Present

We cannot change the cards we are dealt, just how we play the game.

—RANDY PAUSCH, PROFESSOR OF COMPUTER SCIENCE AT CARNEGIE
MELLON UNIVERSITY; AUTHOR OF *THE LAST LECTURE*; DIED OF
PANCREATIC CANCER ON JULY 5, 2008, AT THE AGE OF FORTY-SEVEN

You must live in the present, launch yourself on every wave, find your eternity in each moment. Fools stand on their island of opportunities and look toward another land. There is no other land; there is no other life but this.

—HENRY DAVID THOREAU

A friend once told me, "You are only as happy as your unhappiest child." True. And as a surgeon, I am only as happy as my sickest patient. I worry about my patients before and after surgery, expecting something to go wrong even when I'm sure everything has gone perfectly. After some years of being a surgeon, I assumed that this nervousness would go away, but there is rarely a time when my patients are not in the back of my mind. I scour their charts every morning and night, looking at every vital sign, nursing note, lab value. I text my fellows to get updates, particularly when I notice they've ordered an X-ray or CT scan—and I always notice. The details are critical. A single overlooked lab or test result can lead to a disaster that could have been averted.

There are only two times when I don't worry about my patients. One is when I am in the operating room with them. I'm not cavalier about surgery. I am acutely aware that it is critical to stay focused, keep your mind clear, and not get frustrated. Good surgery is so much about mental toughness and intestinal fortitude in the face of stress and exhaustion and interruptions such as phone calls and pages. And yet the operating room is not a stressful place for me. I am there with my team, my music (generally Tupac on Pandora), and my jokes (which everyone always laughs at, funny or not). I am focused on operating, teaching, and keeping the room light and positive. I love getting to know my patients preoperatively and postoperatively, but none of that plays any role in the OR. It doesn't matter if the patient is nice or a jerk, rich or poor, loving of mankind or a complete racist (like the guy with the swastika tattoo I recently put a kidney into). Most of the patients I evaluate for liver transplants have led difficult lives and made poor choices with regard to their health. I don't judge them for it, though, and I certainly don't claim to understand what their lives were like.

The other time I don't worry about my patients is when they are dead. I feel terrible that they died, and guilty for my role in that, but I have developed a strong coping mechanism that somehow allows me to move on.

I remember most of the patients I've transplanted over the last decade incredibly well, and the details of their operations almost to the last stitch. Sometimes I can't picture their faces or their families, but when I review the operative note, I remember every aspect of the case—how the liver sat, how I struggled on the upper cuff, how I encountered bleeding on the cava. I remember best the ones who had problems—difficult recoveries, lengthy postoperative courses, and returns to the operating room.

This is a sad thing about surgery. Most of our patients do really well, but we spend all our time working on and thinking about the ones who do poorly. One of the things that attracted me to the field of transplantation, other than my love of immunology and my (mistaken) belief that it would impress women at parties, was that it is one of the few remaining areas of surgery where we develop lifelong relationships with our patients. We see them when they are being considered for the transplant waiting list, during their hospital stay, and forever thereafter. This is partially because patients don't want anyone to make decisions about their health other than the doctors who put in their new organs, and also because they are on immuno-suppressive medication, which makes them much more complicated to care for than other patients. With medication suppressing their immune systems, simple illnesses can explode into life-threatening conditions, minor outpatient surgeries become disasters, wounds don't heal, bowel anastomoses fall apart. It is an honor to be a part of their recovery, but also a burden.

Every now and then, I have run across a patient who leaves an indelible impression on me. Jason was one of those people. I first met him in the transplant clinic in 2011. It was a Thursday, and his visit was scheduled for 10:30 a.m. I was running late doing a donor nephrectomy (removal of a kidney from a donor) and had to run to the appointment with Jason still wearing scrubs and a white coat.

Jason was sitting by the door watching me approach. I remember being surprised at how young he looked. (He was thirty at the time.) His parents, sister, and brother were there, too, looking nervous, respectful, and supportive all at the same time. I was struck by the scene because I was so used to seeing older patients, many of them ravaged by alcohol addiction or hepatitis C, and often from troubled families and therefore with little familial support. Overall, with the exception of the unmistakable yellow tinge to the whites of his eyes, Jason looked pretty healthy to me.

We began talking, and I learned that he was a history teacher at a high school in Wisconsin. He particularly loved teaching European history, and had a strong devotion to Scotland, where he had first gone as a college student. In fact, his illness—or, at least, the part of his illness that included his liver—had begun during that summer trip to Scotland. It started with itching, which at first seemed to be no big deal but soon became harder and harder to ignore. He then found himself extremely lethargic, to the point where he had to drag himself out of bed in the morning. Years before (at the tender age of fifteen), he had been diagnosed with Crohn's disease, but that was under control. Something different was going on now.

He fought these feelings for as long as he could, but one particularly bad morning, he dragged himself to the bathroom, looked at his gaunt face in the mirror, and was horrified to see bright yellow eyes staring back at him. He was smart enough to know that this was not a good sign.

As a general rule, turning yellow is almost always bad, and *painless* jaundice, the condition of turning yellow without belly pain, is particularly ominous. It usually means one of two things: either some sort of cancer is blocking your bile duct (for example, pancreatic or bile duct cancer) or you are in liver failure. (If you *do* feel pain, you might be lucky enough to be diagnosed with choledocholithiasis, stones in your bile duct coming from your gall bladder. Although dangerous, choledocholithiasis is typically treatable with the removal of the gall bladder and the placement of a stent in your bile duct—no big deal in comparison to cancer and liver failure.)

Jason saw his primary care doctor later that week—it turns out that when you tell your doctor you've turned yellow, he'll fit you into his otherwise full schedule—where he got a bunch of lab tests, a CT scan, and an MRI. He then saw a hepatologist (liver specialist) the following week. It was there that he found out he had advanced liver disease, caused by primary sclerosing cholangitis (PSC), a rare autoimmune disease in which a patient's own immune system attacks his bile ducts, leading to chronic liver disease, cirrhosis, and the need for liver transplant to survive. He also learned that inflammatory bowel disease (including ulcerative colitis and Crohn's, the condition Jason had) is associated with the development of PSC, and that he was probably going to need a liver transplant someday.

Jason took all this in stride. He was an extremely mellow, levelheaded person. If he was scared, he definitely hid it from his family. Over time, his health stabilized, and he embraced life just as everyone in his family knew he would. He carried his burden with grace and courage, worked his way up to becoming a high school teacher, and continued traveling the world. He even took two high school classes on trips around Europe, acting as their organizer and tour guide. He read voraciously and loved nothing more than inspiring young students to appreciate the world around them.

Unfortunately, his illness didn't really care what kind of guy he was. Jason's energy began to deteriorate. He tried to push through it but, eventually, he could barely drag himself out of bed. His body was swelling up everywhere, and the jaundice had come back. He spent two weeks reading about PSC, liver disease, and liver transplantation, and by the time he made it to my transplant clinic that Thursday morning, he was far more educated on the topic than he had ever hoped to be.

Even though I was almost two hours late, I took my time getting to know Jason. He had such a positive take on life, and he was a natural teacher—something that became obvious when he tried to educate me on Napoléon, Scotland, and the rest of European history I knew so little about. He asked me about my life, what it was like to be a surgeon, what the training was like, what my strategies were for teaching the medical students and residents. I really enjoyed talking with Jason, and I could imagine having him as a friend. (I don't normally relate to my patients in this way, and that's probably for the best.)

I told Jason about the waiting list, the liver transplant operation, the serious risks, the recovery. We discussed the possibility of a "living donation," in which a family member or friend gives the recipient half his or her liver, and of a "deceased donation." Given the severity of Jason's illness, his chances of getting a liver from a deceased donor seemed high. Once all his family's questions were answered, I left the room. My life went on, and I forgot about Jason. His life remained on hold.

Then, a few weeks later, in the dark of night, I got a 2:00 a.m. call from Angela, one of our organ procurement organization (OPO) coordinators (the OPO manages the deceased donors and its members coordinate organ allocation): "We have a liver offer. It's blood type A."

I sat up in my bed. I usually have some sense about who is at the top of each blood type list, particularly when we have a patient with a high MELD score. This particular week, Jason had shown up at an outside hospital with a MELD score of 37, which put him at the top of our list. I knew he would pull one soon.

"Tell me about it," I said to Angela.

"It's a brain-dead male in his sixties," she said. "Died from a stroke."

I had been hoping to get Jason a perfect liver, maybe from a twenty-two-year-old who'd died in a motorcycle accident or from a gunshot wound to the head. Not that a sixty-something-year-old liver is bad—we use livers from donors in their seventies all the time. As long as the pre-implant biopsy looks good, the liver should perform fine. The liver is a miraculous organ, one that "regrows" and heals itself. Still, in general, the older the organ is, the less life it has left in it.

Unfortunately, since I had last seen him, a few weeks before, Jason had become very sick. In fact, he was at the point where, a few days from now, he might no longer even qualify as a candidate. If I skipped him to wait for a younger liver, I might be signing his death warrant. And who knew when the next liver might come? And who was to say it would be better than this one?

One of my mentors, Stuart Knechtle, used to tell me that when it comes to allocating livers, when a recipient's name gets to the top of the list, and a liver becomes available, *that's* their liver. The allocation system is designed to give the next liver to the sickest patient. Besides, plenty of data show that when transplant centers skip over a high-MELD patient, for whatever reason, that patient has a much higher chance of dying while waiting for another organ.

"Okay, Angela," I said. "I'll take it."

I spent the next few hours on and off the phone, speaking first with the transplant coordinator. (The transplant coordinator is different from the OPO coordinator; while the OPO coordinator deals with the deceased donors, the transplant coordinator works with our transplant team to coordinate the recipient side and help take care of the recipients before and after transplantation.) I then called the hospital where Jason was staying and had him transferred to our hospital. Then I called the fellow, Silke, who was working with me on liver transplant, to let her know Jason would be coming in. Then I was back on the line with the transplant coordinator, who updated me on when Jason would arrive. Then back with the OPO coordinator, who told me when the organ procurement would take place. Then the OPO coordinator again, when the procurement got pushed back an hour. This is how it always goes.

I went to see Jason around midday, to talk to him about the upcoming transplant. He smiled when I walked in. His parents and siblings were happy to see me, as he had told them he was hoping I would be his surgeon.

Jason had undergone a few bowel resections to treat his Crohn's disease, and had been on steroids for a while to treat this inflammatory condition. He was thin but healthy—despite his completely failing liver and the recurrent infection in his bile ducts—for one of our patients. I had reviewed his films carefully, examining the cross sections of his liver over and over again, identifying his hepatic artery, his aorta, his shrunken liver, and most important, his portal vein. I traced all the little branches of the portal vein, looking for any large varices that might have formed from his cirrhosis. As I committed his digital anatomy to memory, I could truly picture what these structures would look like when I encountered them in the OR, what tissue planes I would need to get into as I dissected around them with my instruments, where I might encounter side branches and varices. Getting in the right tissue planes is one of the keys to being a good surgeon. The best surgeons always know where to go, and find the spots to dissect where they have minimal bleeding and avoid injuring any structures. Those less gifted are constantly encountering bleeding and other badness. Some of this ability is teachable, some comes with experience. Some surgeons just have it (and some don't). Whether you have it or not, preparation is critical.

Jason's liver did have some decent-size branches off his portal vein, a piece of information I tucked away in my memory bank. His liver was quite big, with the left lobe reaching around the spleen. But it also wasn't the massively swollen organ we normally see in our patients with acute alcoholic hepatitis. I was excited to get the chance to return this young teacher to the classroom.

Operating Room 16, 4:00 a.m.

"Wow. I can't believe how stuck this is!" I told Silke, over the melodious sounds of hip-hop playing on the OR stereo system.

I probably should have expected this—PSC is an inflammatory disease, after all—but I hadn't done a transplant for this illness in a while. I always tell my fellows that liver transplants, and particularly those that occur in the middle of the night (i.e., almost all of them), require a vast amount of internal fortitude in a surgeon. You have to keep your head straight, avoid stress and frustration, stay loose. So much of it is about managing expectations, taking your time, letting things develop.

Overall, in Jason's case, we stuck to the plan—other than throwing in a splenectomy free of charge when, despite our best efforts, a small, inadvertent tear in the capsule that surrounds the spleen wouldn't stop bleeding. After a bit of time spent in the lounge, waiting for the liver to kick in and the bleeding to stop, we returned to the OR, finally ready to close up. We irrigated him and started closing. Everything looked great. I was tired but happy. I looked at the clock: 11:15 a.m. I needed to go talk to his family and then run down to the clinic. I was only three hours late. Not a record.

Jason's postoperative course was a bit . . . complicated. I had to take him back to the operating room a couple of times—first, after his bowel anastomosis broke down where I had rerouted things to plug in his bile duct, and again when his wound came apart. But he handled it with calmness and grace. Despite the complications, he went home a mere three weeks after his transplant, which is a testament to his strength and attitude, and to the support of his family. Over the next few months he dealt with some issues. First, he had to self-administer IV antibiotics at home. Then he suffered a bout of cytomegalovirus (CMV), an infection some of our patients get from the immunosuppression. This required some readmissions and more IV meds. Yet he never appeared down or lost his sense of humor. I saw him a few times in clinic, before he switched over to the hepatologist. After a few months, I stopped worrying about him.

Everything went great for Jason after that—for a few years. He got back to his teaching and was able to make some trips around the country. He enjoyed his family, particularly two nieces born after his liver transplant. But he did have some problems. He suffered from a lot of back pain and osteoporosis (bone disease leading to bone loss), caused by his Crohn's disease and made worse by the immunosuppressive drugs. He broke his legs in a fall, which put him in a wheelchair for a time. Through it all, he remained positive. He kept reading and learning. He got his master's degree, and even started working toward a doctorate in education. Then, for some reason, he started to turn yellow again, and became fatigued. All the original symptoms came back. Something was wrong with his new liver.

We did a workup on him, looking for a recurrence of his PSC, but could find nothing. His ducts looked normal, and the spot where we'd sewn his bile duct to his bowel was unobstructed. All his blood vessels looked perfect. The liver was perfused beautifully. His biopsy showed a lot of scarring and inflammation, but no matter what we tried, we couldn't reverse it. Finally, about four years after his transplant, he ended up back in my clinic for reevaluation. He needed a new liver. (Of course, I was late to his appointment.)

That last appointment with Jason was an interaction I will always remember. We talked about how he had lived his last four years. He didn't complain, even when I asked him about all the complications he had suffered through. He talked more about the good times—the teaching, his students, his love of *Star Wars*, and, of course, Scotland. He'd given me a copy of one of his favorite books, *How the Scots Invented the Modern World: The True Story of How Western Europe's Poorest Nation Created Our World and Everything in It*, by Arthur Herman. I had to admit to him I hadn't read it yet, but I promised I would. I asked him if he had been back to Scotland. He had wanted to go, but the recurrence of his illness had thrown a wrench in that plan. I told him he would get to go after he got another liver. He smiled and said, "Yeah, I hope so." But something about his look told me he knew he wouldn't. Jason had suffered a lot in his young life. He'd been dealt a worse hand than most of us. You wouldn't know it from his attitude, but it was the case. Unfortunately, now the time had come to show his cards. He was at peace with it.

We do a lot of redo liver transplants at our program, but they are much harder. The liver tends to get stuck in there, and blood loss can be extreme. Still, there was no reason to think Jason couldn't get through this operation. Either way, he was prepared.

Not long after our office visit, Jason was called in. A liver was available; it was a good one, a young one. I happened to be out of town at the time, and one of my partners took him to the OR. Sadly, it was one of the bad redos. The team spent hours trying to hack Jason's liver out, and the trouble started early. The bleeding was immense, and his upper cuff was torn. With heroic effort, he did make it out of the OR many hours later, hanging on to life by a thread. Over the next few days, he was on life support, as one organ after another failed.

I got to see him during that period, but in no way did he resemble the person I had seen in clinic just a few weeks before. His body was swollen beyond recognition, with tubes and lines coming out of every orifice. Jason never woke up, dying a couple of weeks after his surgery, with his entire family at his bedside. I hope he didn't suffer.

I wish Jason had gotten a better liver the first time. I wish I had been there for his second transplant, though I don't think for a second that I could have done the job any better. I feel sad that Jason had to be afflicted with Crohn's and PSC—sad in a way that Jason never appeared to be. He seemed so much stronger than I ever would have been. I suppose I should be thankful that he got a few good years after that first liver, years during which he felt entirely healthy. The reality is, although we have come a long way from 1963, when Starzl put in that first liver, we still have a way to go. We have many victories, but the losses are the ones we never forget. They torture us, but also keep us striving to do better.

I took a look at the Scots book and promised myself I would read it sometime, but so far, toasting Jason has had more to do with drinking Scotch. Thanks, Jason, for teaching me a little bit about strength, grace, and living in the moment. In a different world, at a different time, we probably would have been friends. I'm glad I met you.

| 11 |

Lisa and Herb

Should We Do Liver Transplants for Alcoholics?

It is difficult to feel sympathy for those people. It is difficult to regard some bawdy drunk and see them as sick and powerless. It is difficult to suffer the selfishness of a drug addict who will lie to you and steal from you and forgive them and offer them help. Can there be any other disease that renders its victims so unappealing?

—RUSSELL BRAND, *RECOVERY: FREEDOM FROM OUR ADDICTIONS*

Alcohol ruined me financially and morally, broke my heart and the hearts of too many others. Even though it did this to me and it almost killed me and I haven't touched a drop of it in seventeen years, sometimes I wonder if I could get away with drinking some now. I totally subscribe to the notion that alcoholism is a mental illness because thinking like that is clearly insane.

—CRAIG FERGUSON, *AMERICAN ON PURPOSE: THE IMPROBABLE ADVENTURES OF AN UNLIKELY PATRIOT*

"How many of you think we should do liver transplants for alcoholics?"

About half the hands were slowly raised, while the other members of the class looked around nervously. These were third-year medical students, and I was giving my monthly lecture on organ transplantation.

"How many of you think the potential recipient should have six months of absolute sobriety before being offered a transplant?"

This time, the majority raised their hands, and a look of confidence could be seen on most of the students' faces. Obviously, patients should be able to remain sober if they wanted to receive the incredible gift of a liver transplant, the ultimate life-saving resource.

"But what if they won't *live* six months? What if they are going to die in the next two weeks? What if the patient is a thirty-seven-year-old mother of three, or a twenty-six-year-old college graduate who didn't realize the damage he was doing to his liver? Would you ignore the existence of the very young children when you made the decision to let their mother die? Would you stand over the young man, with his parents watching, and tell him you could save him but you've decided he doesn't deserve it?"

I continued: "How many of you think alcoholism is a disease?"

Almost everyone raised his hand.

"What do you think the recurrence rate of this disease is after liver transplantation?"

A few people guessed about 20 percent, which is roughly accurate.

"How many of you think hepatitis C is a disease?"

Everyone.

"And the recurrence rate of that after transplant?"

One hundred percent.

"So, should we transplant in the case of hep C? Couldn't you say the same thing for NASH?" I asked them, referring to the fatty liver disease caused not by alcohol but by obesity, diabetes, and high cholesterol. "Should we not perform transplants in obese people? Or what about patients in renal failure caused by poorly managed diabetes or hypertension?"

In the early days of liver transplantation, saving patients with alcoholic liver disease was generally considered an inappropriate use of this limited resource. Yet now that the practice has been supported by data showing that outcomes for these transplants are as good as or better than outcomes for other diagnoses, the policy has changed. Many programs require candidates to have been abstinent for at least six months. Why? Is a patient who qualifies less likely to go back to drinking after a transplant? And what if a particular patient is so sick from his liver disease that he can't drink? Does waiting that six-month period benefit anyone?

The six-month rule, which has been widely adopted at many transplant centers around the country, came from a retrospective study of forty-three patients who underwent transplant for alcoholic liver disease. In this analysis, abstinence for less than six months prior to transplant was considered a risk factor for recurrence. Multiple further studies have been equivocal on the specific length of abstinence required to reduce recidivism, or return to alcohol use post transplant. To add to the confusion, a recent study from France (where drinking wine is essentially required) showed that well-selected patients with a diagnosis of severe acute alcoholic hepatitis did just as well with transplant and had a similar recurrence rate as those who had abstained for six months.

I have had wonderful success with patients who came into the hospital in acute liver failure days after their last drink, and stunning failures with patients who had not had a drink in years. I remember a twenty-seven-year-old with severe anxiety disorder and acute alcoholic hepatitis, within days or even hours of death, who completely turned his life around after his transplant (and hellish recovery) and is back in school. I remember a young mother who drank on the sly who has rededicated herself to her family and career. I also remember the look of shame and regret on the face of an intelligent, successful, endearing father of three who was admitted to the hospital with a trashed transplanted liver from relapse because he could not free himself from the grips of this deadly disease.

I continue to struggle with this question, whether we should do transplants in alcoholics. Who are the right people to receive this gift of life? In the end, I don't have the answer. But maybe my patients do.

Lisa's Story

I can still remember when I first met Lisa. I had just finished my rounds and decided to swing by her room to quickly deliver my spiel about liver transplant. All I really knew about her was that she was young, forty-one, and sick. Her MELD score was 32, and her disease was alcoholism.

When I walked into Lisa's room, something about her startled me. She had this fresh, young look, a beautiful smile, and although she was sickly and yellow and swollen, there was a sense of joy in her, and a bit of mischief visible beneath the fear and anxiety that accompanies severe illness. Her eyes emitted a hint of sorrow, an understanding of what she had done to get here. Somehow that wasn't what I had been picturing when I heard about this woman with alcoholic cirrhosis who had been abstinent for over a year.

When I got back to my office, I flipped through her chart before writing my note, focusing particularly on the AODA ("alcohol and other drug abuse") assessment that is part of our protocol. The assessment fit with my rapid-fire read of her during our first meeting. Lisa drank wine—usually no more than two glasses a day. She had drunk more when she was younger, but not now. She did think she used alcohol to help with anxiety, some of which may have stemmed from an assault she endured when she was younger. But she had been sober ever since she found out she was ill.

As I read that report, my initial thought was that alcohol had probably played some role in her liver disease, but it may not have been the primary cause. We never really know how much alcohol it takes in any individual to cause cirrhosis. In general, we think that men who have more than two drinks a day and women who have more than one are likely abusing alcohol, but the majority of people who drink at this level will not develop liver disease. Many other things can play a role in the development of cirrhosis, from genetic factors to obesity (leading to fatty liver) to plain old bad luck.

We also know that, when asked about it by a health professional, people tend to underreport how much they drink. So, as a simple rule of thumb, we usually double the amount people report—especially when they are being considered for a liver transplant. Nevertheless, I thought Lisa was at low risk for recurrence, maybe because I instantly liked her. Even I, a transplant surgeon who enjoys drinking alcohol, wanted to believe she didn't really drink that much.

Lisa's surgery was as straightforward as a liver transplant can be. We did find about five liters of pilsner-colored ascites (fluid) in her belly, and a shrunken, cirrhotic liver, which we dissected away from its blood-filled attachments. We stayed in all the right planes, never lost control, never had to turn the music down or stop my constant stream of jokes. When we brought the new liver to the field, we remarked on how beautiful it looked. We hooked up all the blood vessels, and then released the clamps and watched the liver pink up and purr back to life. Shortly thereafter, it began emitting beautiful yellow bile from the duct, and we knew things would be okay. We sewed the ducts together, the donor organ's to the recipient's, took one last look around for any bleeding, and closed Lisa up.

Things went so smoothly that Lisa's breathing tube was removed while she was still on the table. We wheeled her out, victoriously, to the recovery room, and I went down and spoke to her family. Everything had gone great. It was 4:00 p.m. I was even able to get home in time for dinner. Nice day.

Lisa's recovery went well, and when I saw her again in the clinic three weeks after her discharge, her yellow hue was gone and most of the excess fluid had drained from her body. She looked like a "civilian," as I like to say, no longer in the standard-issue hospital gown and slippers. Her smile looked the same to me, and the sadness in her eyes seemed to have gone. I would soon transition her over to my partner in hepatology, Dr. Alex Musat.

When Alex saw her two months later, her liver numbers were perfect. She was pleased with her recovery and was enjoying her family and her life again. He scheduled her to return for another visit in six months—but she didn't show. Then, ten months after her transplant, she was admitted with severe liver dysfunction, her skin as yellow as when she started. Her liver biopsy suggested she had been drinking again.

I went to see her, and there she was, mildly yellow again, back in her standard-issue gown. I awkwardly danced around the issue of alcohol, finally asking her if she had resumed drinking since the transplant. She promised me she hadn't. Her reason for having missed her follow-up visit and lab draws for the last few months, she told me, was that she had been busy. I told her that if she drank again, this new liver would fail.

Of course, none of us believed her. Unfortunately, we had seen this before. Over the next few years, Lisa was in and out of the hospital with serious liver dysfunction. For a while, she continued to deny her drinking, and then eventually admitted to drinking just a little bit.

Not even five years after her transplant, I got an email notifying me that she had died. I knew that her liver was shot, that things couldn't have ended any other way. Yet her death stayed with me. I could still see her smile, still remember her young family, her children. Why couldn't it have turned out differently? What did we miss? I consoled myself with the idea that at least her family had gotten to enjoy her for a few more years, that in some way the gift of life must have been worth it. But was it?

More than three years after her death, I reached out to Lisa's husband, Jay, to see if he could help me understand what had happened, if we could have done something more to help his late wife. Jay struggled with this request. He and his three children were healing, moving on, and he didn't want to pull the scab off the deep wounds they had endured. He also admitted that he had anger toward us—he couldn't understand how we could give Lisa a new liver and yet not treat her alcoholism. To him it was like putting a "Band-Aid on a gushing wound." Yet he ultimately decided that if Lisa's legacy could help someone else, help us understand and talk about alcoholism and mental illness, a meeting would be worthwhile.

Jay met Lisa right as he was finishing college. She was well read, well traveled, and beautiful, and she quickly became his best friend. His career was taking off beyond his wildest dreams, and everything seemed to be going his way, especially now that he had found Lisa. In retrospect, he admitted that there were some potential warning signs. He had known about her strained relationship with her parents, although he had barely met them. He knew she'd had a "rough" upbringing and that, when she was around sixteen, she sat her mom down and told her, "Either you kick Dad out and get a divorce, or I'm leaving." Jay thought it likely that Lisa's father was himself an alcoholic, or at least someone who abused alcohol. And "it sounded like he was probably verbally if not physically abusive," he told me. Jay was surprised at how little contact Lisa's remaining family had with her father, and with Lisa, especially once she and Jay had children. The isolation from her family was devastating for her.

Did this isolation play a role in Lisa's alcohol use? Jay thought it did, but then he brought up something else: "I think the root cause was the PTSD, and I think the PTSD came from something that happened in college that was very violent."

Lisa was a victim of sexual assault, and never really got over it. Jay blames himself for not understanding how much it affected her, and how she never really dealt with it. "Frankly, early on in our marriage, just because I was so young, I wasn't ready for that news," he told me. If he'd realized how much it was still bothering her, he added, "I would have said, 'Hey, you need to seek out some counseling.'"

The signs of alcohol abuse were gradual, and easy to miss. There were a few occasions when Jay would find an empty beer can under the sink and ask Lisa about it. She would simply remark that she'd been cleaning and forgot about it. As Jay told me, "In marriage, you've made vows, and so you want to be trusting. And so I would dismiss it . . . but there were inklings that there might have been issues."

Each episode seemed to have an explanation, a story that didn't involve Lisa being in a drunken stupor. Eventually, her problem became undeniable, but they both continued to try to ignore it. In this way, Jay and Lisa were able to avoid facing her condition for years, something that became next to impossible when she woke up yellow and was diagnosed with advanced cirrhosis, most likely due to alcohol abuse. Ultimately, Lisa's health deteriorated more, and she made her way to our hospital for her transplant.

It is hard for me to write about what the next four years were like for Jay, Lisa, and their family. As I flipped through Lisa's file, reading about the many phone calls, clinic visits, outside hospital admissions, and transfers to our institution documented in her medical record, I could only imagine the confusion, fear, and, ultimately, desperation she and her family felt. And then I read the predictable, heartbreaking denouement at the end of her record. There were a couple of short notes from me—reminding me when I'd stopped by and said hello and when, on the way out the door, almost as an aside, I'd said, "Lisa, you really shouldn't drink anymore. It will kill you"—as if, somehow, that would be enough, as if I had done my part. What I couldn't glean from the notes was how, every day, from a few months after the transplant until Lisa's death, maybe fifteen hundred days or more, Jay and his family struggled with this disease.

There were some trips to rehab, but they never worked. Over a short period of time, her transplanted liver began to fail. Her readmissions started increasing. Countless times, Jay described finding her unconscious, not knowing whether it was from alcohol or a failing liver. He'd call an ambulance, and she would get admitted for a few days, or weeks, and then she'd come back home and repeat.

Eventually it became obvious to Jay and his family that Lisa was not going to be cured. She was back where she'd started, just as yellow as before and totally confused. The number of times she was in and out of emergency rooms, hospitals, and rehab centers is astounding. Near the end, she was seen by the palliative care service and, ultimately, hospice. In her ever-decreasing lucid moments, she continued to deny the effect alcohol was having on her life and health. Finally, just three weeks before her death, she apologized to Jay.

Why did it take Lisa so long to admit what alcohol had done to her and her entire family? Jay thinks it was because of embarrassment, because of the stigma attached to alcoholism, and mental illness in general. She felt it was something she should have been able to deal with on her own, without anyone else knowing how much she was struggling.

Lisa died at the age of forty-five. I'd like to say she died at home, surrounded by family, having finally understood her disease and reached a state of peace that gave those who loved her some comfort as she took her final breath. But it wasn't like that. She ended up in the ICU, on a respirator, with tubes and lines poking into her. Maybe that's what she wanted—maybe fighting to the end, hoping for more time with her family, was worth it to her. But as I review the details of her posttransplant life, I find it hard to accept the hell Jay and her family were put through.

Lisa didn't die of liver disease; she died of mental illness. She was addicted to alcohol, which she'd likely turned to in response to anxiety, unrecognized PTSD, and a genetic predisposition to addiction. When we put a new liver in her, this simply reset the clock. It didn't do anything to treat her disease. In some ways, this is a microcosm of how our whole health care system works. We celebrate, and pay for, the big, sexy interventions—the operation, the cardiac cathe-terization, the heroic treatment that is technically challenging and potentially risky. But what really matters, and yet what our health care system doesn't prioritize, is the day-to-day caring for chronic disease, the incremental, preventative care that can avert transplant altogether. Alcoholism is never actually cured. It can be managed, it can go into remission, but it is always there.

So, should we have transplanted Lisa? I don't regret that deci-sion, but I do regret how we managed her afterward. We knew she was high risk, and we knew that about 20 percent of patients re-lapse after transplant. She had so much to live for, was charismatic, intelligent—and believable when she claimed she'd stopped drink-ing. And because she initially did so well, we mistakenly thought she would be okay.

We try very hard at our program to provide support and coun-seling, to hook patients up with mental health professionals upon their discharge, or to send them to an appropriate rehab facility if we think that is needed. But when Lisa told us she wasn't drinking anymore, we fell for it, just as Lisa did. She was very convincing because she herself was convinced she was okay.

I still feel horrible about this case today. Jay and his family didn't deserve this. Neither did Lisa. She was a good person with a bad disease.

Yes, addiction is a disease. Having an addiction doesn't mean you are weak or bad, or that you deserve to die. Addiction shouldn't be an embarrassment, but you need to ask for help. Lisa was just too embarrassed to do the very thing that might have saved her life.

Still, I ask myself, if she had asked for help, would we have known what to do?

Herb's Story

Cunning, baffling, and powerful—that is how Alcoholics Anonymous describes the allure of alcohol. That description conjures images of a serpent slithering around, its tongue lashing out, waiting to strike at will, and the victim powerless to resist. The reality is that some people have no problem drinking casually, while others seem to fall under the grip of this cunning force. Why? What causes seemingly thoughtful, intelligent people to become addicted to alcohol, to the point where it destroys their livers and their lives?

No doubt, there is a genetic component, but that isn't the whole story. I've met many patients with no family history of alcoholism, and no other form of addiction, no specific reason they would need or want alcohol to deal with their daily lives. But just as many people associate coffee with starting their day, many people associate alcohol with relaxing at the end of it. Pretty soon, though, they start taking a drink earlier and earlier in the day. And once they're hiding it from their families, they're too far gone. The disease is that cunning.

HERBERT HENEMAN, DICKSON-BASCOM professor (emeritus) of management and human resources in the School of Business at the University of Wisconsin–Madison, was not your typical corner-of-the-dive-bar alcoholic. Herb grew up in St. Paul, Minnesota. His father was a business school professor, and his mother was a stay-at-home mom. Herb had a happy childhood and a very supportive family. He describes his parents as somewhat heavy drinkers, particularly his father, but he remembers no health issues, legal problems, or family crises related to alcohol. His parents would not let him drink while he was growing up, and although he did drink a small amount with his friends in high school, he does not consider it excessive when he reflects upon it. Still, he did grow up in a culture of drinking and was used to seeing alcohol at meals and other social gatherings.

He attended a small liberal arts college, where he excelled. He occasionally did some "hard drinking," on weekends, at parties, but didn't drink during the week. Once he finished college, he began graduate school at the University of Wisconsin and married his high school sweetheart. Everything was falling into place.

Herb doesn't remember a specific time when he suddenly increased his drinking, but alcohol slowly started to play a bigger role in his life. He continued to excel at work, easily getting promoted and thriving within the university's tenure structure. But at some point, he remembers thinking about alcohol more and more. He found himself stocking his liquor cabinet more frequently, and he started drinking during the week. Then he found himself drinking during the day, and not just at home.

He doesn't remember a dramatic liftoff. Alcohol just slowly started to permeate everything he did. He started hiding his drinking from his wife and kids, and drinking alone. He found himself getting sick more often, episodes he described as the flu, or exhaustion, or general weakness. He'd be on his back for three to five days, but once he felt better, he'd say, "Good. I'm better. Now I can drink again." Eventually his physician told him it was his drinking that was causing his health problems, but Herb didn't buy it. He knew he could stop if he wanted to.

Things came to a head for him on Labor Day 1990. He and his

wife were hosting a wedding party for his niece. Herb had told his wife that he'd stopped drinking, but, as he says candidly, "I hadn't." As he remembers that day, "I had in my mind that I really didn't want to be a part of that party, and so I was drinking very heavily prior to it." Something nobody knew.

Herb was hammered—stumbling, sweating, and generally not looking well. A nurse who happened to be at the party thought he was having a heart attack, and an ambulance was called. En route to the hospital, a blood-alcohol level was taken, and it came back at 0.375. Herb was placed in detox, and from there he went to a twenty-eight-day inpatient rehab program. Once he sobered up, he realized he was sick. His liver hurt and felt swollen (probably alcoholic hepatitis), his thinking was impaired, and his digestion wasn't right. He was sober for the entire twenty-eight days, and told everyone there he was committed to sobriety, that he had it beaten.

He relapsed the first day out. About two weeks later, his wife caught him drinking. She was devastated. Herb went to detox again, and once he was sober, he agreed to be committed to a behavioral addiction unit as an inpatient in a psychiatric hospital in Milwaukee. How far he had fallen! Still, this wasn't just any inpatient program. It was the McBride Center for the Professional, a branch of the Milwaukee Psychiatric Hospital and a place where patients with successful careers can be treated for their addictions. Being surrounded by other professionals was absolutely critical for Herb. In a regular rehab facility, it is too easy for people of Herb's social status to look at the people around them who are also struggling with addiction, especially if they are from a different walk of life, and say, "I'm not like them. I can control this." Even so, at McBride, Herb resisted. He didn't really want to participate in the group sessions; he just wanted to read about alcoholism and beat this thing using his own brain.

"On a Sunday morning," he told me, "I went to a little interdenominational church service that was being held in the hospital. And as I walked in the door, this woman began playing 'Amazing Grace' on the piano. And that was my true turning point. All of a sudden, and I cannot explain it to you in rational terms, but I began to feel that amazing grace. And I sat during that whole service just crying and coming to that full realization that truly I had a problem." Herb spent three months as an inpatient.

Although he was finally owning his alcoholism, he was also facing the fact that his liver was failing. He was diagnosed with cirrhosis. After three months, he was released from the psychiatric hospital and found himself in the office of Münci Kalayoğlu, who had trained in transplant under Thomas Starzl and started our program in 1983. Münci told Herb that if he was able to stay sober for a year, he would perform his transplant. But, he told Herb, "there is one thing I want you to understand. If you ever, after your transplant, go out and drink again, I'm going to come over to your house with my pocketknife and take back the liver."

Herb has grasped the fact that his alcoholism will never be "cured." It is always lurking, ready to come back with a vengeance. "The other thing that really helps keep me sober is that I was so fortunate to receive a transplant, particularly back then. It would be an absolute dishonor to my donor family for me to go out and drink again and somehow do any damage to my liver."

After his transplant, Herb had one brief readmission for a rejection episode. Otherwise, he has had no problems with his transplanted liver for more than twenty-five years. I asked him how the transplant changed his outlook on life.

"I think it changed it much for the better. Between that and being a recovering alcoholic, there's an appreciation of the sweetness of life that i never had before. I have a much, much greater empathy for other people. I try to be of service to other people in any number of capacities, both in the transplant community and in the alcoholism community. So, I've led a much fuller life than I otherwise would have."

I asked for his thoughts on transplanting patients with acute alcoholic hepatitis, patients who clearly can't survive a waiting period of sobriety prior to transplant. He said, "My own experience was very much that recovery needs to be a very serious, lifelong commitment, one day at a time, and that people who try to go it alone are not very likely to succeed."

I agree. A liver transplant is a necessary tool that treats one symptom of alcoholic liver disease—it replaces the dysfunctional organ involved. But it can't be looked at as a wake-up call, a treatment in itself for the disease of alcoholism. Alcoholism demands lifelong support, counseling, and mental health treatment. It is our job (and the job of our colleagues in mental health) to identify those patients who have insight into their disease, and some chance of success after transplant. Otherwise, we are likely to fail with the limited resource of a new organ. For those who are too sick to wait, we should require and provide intensive counseling and mental health treatment until that insight is achieved. We also can't look at relapse as a failure, and turn our backs on the person. When a transplant patient relapses, it is often seen as an affront to us, the donor, and the entire transplant process. That is the wrong way to look at it. Alcoholic relapse needs to be managed like relapse in the case of any potentially treatable disease.

I talked to Herb about Lisa, and her story made him sad. We both agreed that she was never able to develop insight into her disease until the very end. We also agreed that she probably looked at others struggling with alcoholism as different from her. He mentioned that he was part of a new hospital committee trying to develop a program to follow patients after liver transplantation for alcoholism, particularly high-risk patients. That gives me some hope.

| 12 |

Nate

The Selection Meeting,
or Who Gets an Organ and Why?

Who shall live and who shall die, who in good time, and who by an untimely death, who by water and who by fire, who by sword and who by wild beast, who by famine and who by thirst, who by earthquake and who by plague, who by strangulation and who by lapidation, who shall have rest and who wander, who shall be at peace and who pursued, who shall be serene and who tormented, who shall become impoverished and who wealthy, who shall be debased, and who exalted. But repentance, prayer and righteousness avert the severity of the decree.

—"UNETANAH TOKEF," POEM READ AT ROSH HASHANAH AND
YOM KIPPUR, AUTHOR UNKNOWN

To hope under the most extreme circumstances is an act of defiance that permits a person to live his life on his own terms. It is part of the human spirit to endure and give a miracle a chance to happen.

—JEROME GROOPMAN, *THE ANATOMY OF HOPE*

We have liver selection every Wednesday at 1:00 p.m. The key participants include the transplant surgeons, hepatologists, radiologists, fellows, residents, social workers, some of the floor nurses, and our many transplant coordinators. The last are the people who interact with the potential transplant candidates the most during the evaluation, summarize their stories for us, help them get their tests done, and communicate with them day and night. A separate group of coordinators takes care of these patients pre- and post-transplant. The surgeons get a lot of credit, but really, it is these people who make everything happen.

At these meetings, we are given a list of patients being considered for transplant. Usually there are about ten to consider, and each is listed by name, age, BMI (body mass index), diagnosis, and MELD score. One particular Wednesday, I was running the meeting, standing in for our chief of liver transplant. Most of our agenda that day included patients with MELD scores above 20, and a few above 35. Without knowing anything more about them, we knew that these patients had very dysfunctional livers and were at great risk of dying from their disease. As I scanned the list, I noticed with discomfort that the diagnosis codes for these patients were almost all the same: the majority had ALC, or alcoholic liver disease, with an occasional NASH, or nonalcoholic steatohepatitis, thrown in. (Since better treatments for hepatitis C are now available, this was not surprising. With the number of patients being transplanted for hepatitis C dropping, alcoholic liver disease has become the most common indication for liver transplant, with NASH rising in incidence.) What was making me uncomfortable was that, the whole time we were discussing these patients with high MELD scores and self-induced liver disease, sitting behind me was Nate. I couldn't help wondering what he was thinking.

DURING HIGH SCHOOL in Des Moines, Nate thought becoming a paramedic would be cool. No one in his family was in the field of medicine. At that point, he wasn't considering a career as a physician, but he liked the idea of becoming an expert in a field where he could help people. He decided to take a year off before college to train as an EMT—but then he began losing weight.

The weight loss seemed to be a result both of not really having an appetite and of the almost constant diarrhea he had developed. At first, he thought it was nothing major—maybe a bug or some sort of food intolerance. But the symptoms persisted, and despite his best efforts, his weight fell from a healthy 180 pounds to 129. He finally made his way to a local doctor, who ordered a colonoscopy and diagnosed ulcerative colitis. Nate was hospitalized for about a week, placed on steroids, and forced to endure a barrage of tests and procedures. The doctors put him on oral medicine, and his symptoms got better. No one talked to him about what his illness could mean to his future, whether surgery would be needed, what other disease processes could be involved. His symptoms were better, and life went on.

The following year, he enrolled at Drake University, in Des Moines. He lived in the dorms, studied politics, and continued working as a paramedic. By the end of his freshman year, he realized that the idea of being a doctor appealed to him more than that of being a paramedic. As a sophomore, he added premedical courses to his curriculum. He knew it would be a lot of work, as the requirements were daunting—biology, chemistry, organic chemistry, calculus, physics— but it seemed an exciting career to him, fulfilling.

Then, in January of his sophomore year, everything changed. He was at a routine GI appointment and his doctor decided to check some labs. Nate got a call shortly after and was told that his LFTs (liver function tests) were all elevated (which is usually a sign of injury to the liver or the bile ducts in the liver). "Probably nothing," the doctor said; "maybe a reaction to your medication." Nate got a CT scan, and was told things looked okay, but that he should get an MRCP (a type of MRI in which doctors look at your bile ducts). He did so, but still thought nothing of it. He was twenty years old, felt fine, and wasn't jaundiced.

The following Monday evening, he got a phone call that is a textbook example of how *not* to deliver bad news:

"Hi, Nate. Are you driving right now?"

"Umm, no," Nate said.

"Are you at home?"

"Yes."

"Okay. I'm really sorry to tell you that you have this disease, primary sclerosing cholangitis, or PSC, and you're going to need a liver transplant in the next few months. I'm going to send you to the University of Iowa."

"What? Okay . . ."

Nate came to the conclusion that he had just been given a death sentence. He turned off all the lights, put on his favorite album (by the Red Hot Chili Peppers), and cried. For an hour.

In hindsight, he admits that he should have known something was gravely wrong. He was very itchy (one of the most debilitating symptoms of PSC), but he attributed it to some sort of allergy. And he was tired, but he figured that was from many late nights out with friends or studying. With his PSC diagnosis, he finally made his way to the University of Iowa and saw a hepatologist, the first medical professional he talked to face-to-face since he got his MRCP. And this doctor looked over his labs and film reports and told him he was fine—for now. Nate had no idea that the average time from diagnosis of PSC to needing a liver transplant is ten years.

The following summer, Nate was back at school taking organic chemistry. He really hadn't thought much about the PSC since his diagnosis and was busy enjoying college and getting more focused on the idea of a career in medicine. But he was still itchy all over. Also, he found it painful to walk, and he would wake up in the middle of the night from either the itching or the searing pain he'd caused from scratching the skin off his arms, legs, and forehead. After some more searching online, he started taking oatmeal baths or sitz baths every night—for hours, just to try to soothe the agony. He knew he needed help. He made an appointment at the Mayo Clinic. There he learned about his risk of cholangiocarcinoma (cancer of the bile duct), cholangitis (infection), and cirrhosis. He was told he needed an ERCP (endoscopic retrograde cholangiopancreaticogram), a procedure in which a GI doctor puts a scope down your mouth, through your esophagus, stomach, and duodenum; cannulates your bile duct; squirts contrast in to see if there are any blockages; and, if needed, places a stent to promote drainage, which helps with the itching.

Over the next year, his twenty-first, Nate had many ERCPs, all at the Mayo Clinic. Each one seemed to improve how he felt, but in between, his itching would return. On top of that, he noticed he was turning yellow, losing weight, and never hungry. This was becoming his new normal.

Nate adapted to this amazingly well. He developed an interest in travel, one that has stayed with him his whole life. Over the years, he went backpacking through Europe and China, and to Finland and Russia by boat. He knew he was different from his friends and family members, and that he couldn't "beat" his chronic illness. But his motto was that "if PSC was going to shorten my life, I wasn't going to shorten it any further myself." He still remembers his last alcoholic drink: it was a Guinness he poured himself in a Guinness factory in Dublin.

Three years ago, Nate began medical school at the University of Wisconsin. A career in medicine is the ultimate example of delayed gratification, which is a particular challenge for someone with a chronic disease associated with so much uncertainty about one's long-term health or even survival. Nate is trying to live his life, but at the same time, he wonders about his death. He has generally avoided sharing this with his med school classmates, even though all of them are entering the business of taking care of sick people. Once again, illness finds a way to separate the afflicted from the people around them.

Nate recently finished his surgery rotation and, shockingly, wants to be a surgeon. The training is brutal and interminable, the decision making is challenging and stressful, the failures are crushing, the guilt can be intolerable, but this young man clearly has been bitten by the surgery bug. Nothing else will do for him, even with his unique challenges.

BACK IN THE liver selection meeting, we started talking about a woman in her late thirties with alcoholic liver disease. She had cirrhosis, ascites, jaundice, encephalopathy, and a MELD score of 40. She was close to death. But she also had a supportive family, had insight into her illness, and was committed to turning things around. We decided to list her.

And there was Nate, trying like crazy not to crawl out of his own skin or scratch himself like a madman during the selection meeting, barely able to concentrate, while we were working through a list of high-MELD patients, all of whom had trashed their own livers, and all of whom would be way above him on the transplant list. There he was, working his ass off to get through medical school, so he could learn to help people with illness, but the one thing he likely needed to achieve his goal of becoming a surgeon was a liver transplant. And these people were taking all the good ones.

Yet Nate doesn't look at it that way. He knows that many of these patients are sicker than he is. He doesn't judge them in the way so many of us do. He wants them to get better. He told me, "They deserve the transplants more than me." He also told me that he has always believed, "maybe naïvely, that if I need a liver, I will get one." He also feels lucky that, unlike many of the patients he takes care of, he has an incredibly supportive family and very close friends.

Nate really has two options. Option one is to sign up for a living-donor liver—both his sister and his brother have offered and are in the process of getting worked up. Nate is definitely struggling with the idea of putting them through such a big surgery, one that is filled with risks and uncertainty. Option two is to wait on the list. The good news is he has just been awarded a MELD exception. This means that we wrote a narrative about him, discussing his horrible pruritus (itchy skin), his inability to continue with his job, and his frequent admissions to the hospital. The regional review board, which consists of surgeons and hepatologists from all the transplant programs in our region, agreed to grant him a 22. This means that a majority of liver specialists in the region agrees that the MELD system does not adequately represent Nate's mortality (or need for a liver) and is willing to have him move up the line to receive a liver in the region. Every three months, we will rewrite the narrative to try to increase the score, and we hope that over a year or so his score will get high enough to pull a liver for him. But there are no guarantees. The board could vote no at any time, and his score could drop down to his physiologic MELD, that is, closer to 15.

Nate's story beautifully illustrates the challenges inherent in the allocation of limited resources. I have no problem transplanting livers into alcoholics who have not demonstrated six months of abstinence, if they have good support, insight into their disease, and a commitment to change. We get burned occasionally, but we also have great successes. I also have to believe that we as a society *want* to transplant people such as Nate, and not make him suffer for so long on a list while he risks losing the opportunity to go to medical school, or travel, or live his life. There has to be a better way.

One idea would be to automatically give a large MELD exception, perhaps even a MELD score of 40, to the 8 percent of people on the transplant list with biliary diseases such as PSC, PBC, and other rare primary liver diseases that lead to a miserable existence but low MELD scores. For these patients, liver transplant is the only effective way to halt the progression of their disease, so perhaps they should be moved to the top of the waiting list. For secondary diseases of the liver caused by alcohol, fatty liver, and hepatitis C, where preventative, incremental care is the desired treatment and where liver transplantation is just a resetting of the clock, the MELD system could be applied without limiting access to the patients with primary liver disease. (Again, the livers used by the small percentage of patients with primary liver disease would leave plenty of livers to go around for these other illnesses.) Perhaps this would help shift the focus of management to prevention for secondary diseases of the liver rather than thinking of transplantation as a reasonable option when it could potentially have been prevented. Still, if we develop a system of allocation in which we judge one disease different from another, are we going against an ethical norm not to judge people for how they acquired their illness?

Another option is to get away from an allocation system based entirely on a person's risk of dying on the waiting list. Perhaps we could somehow incorporate quality of life, age, number of hospital admissions, loss of work, and likely outcome, including likelihood of returning to some level of productivity, into our decision to transplant an organ in someone. For now, though, the system is not about to change.

Nate was thrilled that he'd gotten his MELD exception. His plan was to finish his third year of medical school and then do his fourth over two years. He hoped that, during that time, his MELD exception score would continue to climb, and he could get a liver transplant, recover, avoid complications, and move on with his life—which he hoped would include a surgical residency (and maybe a transplant fellowship). If, at some point in the next year, his incremental bump in MELD was not approved (which is quite possible), he would try to move forward with a living-donor liver from someone in his family.

ALMOST TWO YEARS had passed since I first met Nate and heard his story. Since then, I'd watched him valiantly finish his third and fourth year of medical school, and despite my mild protestations, he'd held on to his desire to become a surgeon. He never once asked, "Why me?" or spoke about how unfair it was that he was unable to obtain a deceased-donor liver.

Then, amazingly, a cousin of his, a young, healthy woman he barely knew, stepped forward to offer him half her liver. A few weeks before the surgery, Nate asked me if I would be his surgeon; I was flattered and at the same time hesitant. I really cared for Nate, and felt like a mentor to him, but I wondered if I would be able to be aggressive enough, to take the (calculated) risks and make the split-second decisions I might need to during the surgery, or would I hesitate, filled with thoughts of what a complication might mean to his future?

Before I could answer, Nate threw in this little teaser. "Oh, by the way, my class chose me as the graduation speaker! I'm going to talk about my experiences with liver disease and the transplant. The graduation is seven weeks after the transplant. I should be ready by then, right?"

I CHIPPED MY way around Nate's massive liver, painstakingly working on one side and then the other, making a little progress here, a little there. His liver was so enlarged from his PSC, it took up almost his entire abdomen. I was making decent headway until the end, when I pulled just a little too hard, and blood erupted. For a brief second, the thought *I can't believe I just killed Nate* actually ran through my head—and then I pulled myself back into surgeon mode. I had options. I had been smart, had made my way around his cava below and above as early as I could in the operation, just in case this happened. Now I kept pressure on the tear in the hepatic vein with my hands and let anesthesia know my plans. As I looked up, I noticed Nate's blood pressure on the monitor: fifty-four over thirty. Not good. Anesthesia poured in fluids, and I placed the lower clamp on the vena cava and then slid the large Klintmalm clamp around the upper cava. All the bleeding stopped. I carefully cut out the liver, taking care not to cut the cava out with it. The timing was perfect: the new liver, then being prepped on the back table, was ready for implantation.

Luis and Yucel, two of my partners, scrubbed in at this point, and we brought the new liver onto the field. They plopped it down in the cavity, where it took up less than a quarter of the space Nate's old liver had required. We sewed it in, and everything went perfectly. As I watched the liver go from its pale brown color to pink, and saw bile dripping out of the duct, I felt such a sense of joy. Nate's hell was almost over.

I went down to talk to his family, pictures of Nate's old liver, and even the new one, in hand. They were so thrilled they erupted in cheers. I felt like a conquering hero. This is one of the best parts of being a transplant surgeon. Transplant patients usually have lots of family around, and there is nothing better than seeing them explode with happiness when you tell them how smoothly everything is going.

The next few days went well for Nate. He woke up, had his breathing tube removed, and came off the medicines that were supporting his blood pressure. He looked tired but also thrilled to be through the operation. His ultrasound looked great—all the blood vessels were wide open—and his liver numbers were normalizing. There was one weird thing, though: He had a fever when he hit the ICU after surgery, which in itself is not that abnormal. But blood cultures taken during that fever very quickly grew out fungus. Hmm. Nate's gigantic liver, with all the blockages of his bile duct, must have been filled with that nasty bug. We started him on antifungals, and held our breaths.

Post-op day five, Sunday morning: I noticed that Nate's labs were up a bit. Not horribly, but definitely up. It was probably nothing. I touched base with one of my fellows, who had already ordered an ultrasound. I checked the computer multiple times over the next hour, waiting for the films to come up, but nothing appeared. I was annoyed that the ultrasound hadn't happened yet, but at the same time I felt confident that it wouldn't show anything abnormal. Back at home, I tried to hang out with my kids, but my mind kept wandering back to Nate.

Then I felt my phone vibrating in my pocket, announcing the arrival of a new text. I pulled it out and took a look. It was from Nate.

"Dude, my artery's out," he said. That had to be the first time a patient had let me know something like this. Nate was watching while they did the ultrasound, and had realized they couldn't find the artery to his liver. It must have clotted off.

My stomach jumped into my mouth. I knew immediately what this meant for Nate: he was screwed. As I made my way into the hospital, thoughts of all the nasty complications that awaited him swirled through my head. By the time I got there, he was already in the operating room.

Sure enough, the hepatic artery that we had sewn together from his own hepatic artery to the donor liver artery was clotted. The artery is the main supply to the bile ducts in the liver, and when it clots off, it almost always ruins the liver. We cut out the sutures that we had placed to sew the artery together and pulled a bunch of clot out of the lumen. We sent the clot and a rim of arterial tissue at the anastomosis site for a culture. We then cleaned the artery out with some catheters, squirted in some clot buster medication, and sewed it back up. Then we started infusing some blood thinner through his IV. As I watched the blood thinner drip into his veins, I knew what would come next: bleeding.

One other thing we did was relist Nate for another liver. For patients who have an artery clot in the first two weeks after liver transplant, the United Network for Organ Sharing, or UNOS, will allow them to be listed at a MELD of 40, meaning they will be at the top of the list for a new liver. The challenge now was, if the artery stayed open, should we retransplant, or see how things went?

I immediately wanted to retransplant. We'd gotten away without doing so in the past, but wouldn't a new liver give Nate his best shot at becoming a surgical resident? Was I thinking clearly, or was I conflicted because of my relationship with him? I wasn't sure.

Nate's artery stayed open, but it turned out to be the least of his problems. Over the next few days, three things happened. He started bleeding, his liver function tests started to go up, and his intraoperative cultures grew out more fungus.

I decided to take him back to the OR to wash out the fungus, get a liver biopsy, see what was bleeding, and look at his artery. Nate was not thrilled about this, but he felt like shit. He asked me if he was going to be okay. I hoped so.

In the OR, the liver looked pretty good, actually. I cleaned out some blood and grungy tissue and did a biopsy. But here's the kicker: his bowels were so swollen that I couldn't get him closed. I packed his abdomen with sponges and a suction device and brought him back to the ICU sedated and intubated. What followed was one of the worst weeks of my professional life, and no picnic for Nate.

We took him back to the OR every two days for repeat washouts and attempts at closure. In the meantime, his biopsy came back showing severe rejection and, at the same time, his new intraoperative cultures showed fungus and a variety of other drug-resistant bacteria.

I kept trying to put on a positive face for Nate's family, but I really didn't know how things would play out. We were trying to walk a tightrope, increasing his immunosuppressive meds at the same time that we increased his antibiotics. We walked the same tightrope with his blood thinner, trying to find a dose at which his blood would be a bit thinned but at which he wouldn't bleed.

Finally, on the fourth try in the OR, we got him closed. The breathing tube came out the next day. I expected the first words out of Nate's mouth to be "What the fuck did you do?" but he was too tired and weak to utter such a long sentence. Thankfully.

I would like to be able to say that everything went smoothly after that. But that wouldn't be entirely truthful. Somehow Nate hung in there, despite numerous procedures, consults, needle sticks, scans, biopsies—you name it, he had it done. I joked with him that he would be the most experienced intern who ever existed, but I did start to wonder if it was realistic to think he'd even make it out of the hospital.

After a few weeks, Nate went home for about two days, but then came back sicker than ever. More tubes and drains were stuck in. Now we had a new problem: bile was leaking out of the ducts that drained his liver into his bowel. We put a tube through his liver, through one of the ducts, and into the bowel, and drained his bile into a bag he would carry around. He hung in there, somehow. He looked like a skeleton, with sunken, yellow eyes. As his graduation day approached, we all wondered if he would make it to the dais.

Union South Varsity Hall, May 12, Medical School Graduation

I felt nervous as I looked at the lineup of medical students in their caps and gowns, waiting to march into the hall to receive their diplomas. It didn't take me very long to find Nate—he looked pale green and shaky, and was the only one using a walker. I wanted to believe he was just nervous, but I knew that wasn't the case. I went to give him a hug and could feel his ribs beneath my hands. I asked how he was feeling, but I didn't really hear his answer. I asked if he was ready to deliver his speech. He said he thought so. His wife, Ann, had a copy of his speech, which she'd finish delivering in case he collapsed in the middle.

Once they all marched in, I went to the back of the hall. There must have been a thousand people in there, and I could just make out Nate, sitting onstage to the dean's left. I watched the dean say a few words and a few other speakers give their words of advice, words I have heard so many times—be nice to the nurses, bring cookies; the patients come first; blah, blah, blah. It was all good advice, but so predictable. Then it was Nate's turn.

He slowly walked over to the lectern, looking quite unsteady. Even though I knew he was weak, and I could see his hands shaking as he held a copy of his speech, when he began speaking, his voice sounded surprisingly strong and confident.

Then Nate gave one of the most inspiring graduation speeches I have ever heard, one filled with the lessons he'd learned during his illness. Despite everything he had gone through, he focused on the importance of hope.

Hope, he said, "may be the most powerful currency we carry as doctors." It was hope, he said, that had allowed him to survive for two years on the waiting list, hope that had let him take that leap of faith and undergo a liver transplant at such a young age. His illness had separated him from his classmates—sure, he'd finished med school, but his experience had been so different from theirs. When they went home to study, or out to the pub to get a drink after class, he would sneak off to the third floor of the hospital to get hooked up for a plasmapheresis session, in which his blood was drawn out his arm through big needles and cleansed in a machine. He had spent the majority of the last two months in the hospital, just as his classmates had, but his time had been spent lying in a hospital bed, or in the OR, or in procedure rooms. While his classmates had recently enjoyed the match process, in which they opened envelopes and found out where they would spend the next three to five years of their lives in their training programs, he had simply hoped he would be alive for that long.

"Hope comes in many different forms," he said. "For some, it's simple. Hoping that their cough goes away or isn't the harbinger of something more serious. Hoping that their pain is relieved. Or just hoping that their physician will be compassionate and understanding. For other patients, their hopes loom much larger. It's hoping for a cure. It's hoping for more time with family and friends. Or, for me, hoping that transplant would offer my wife and me a chance at a more 'normal,' healthy life."

He continued: "Your patients will be looking to you to offer hope in their darkest moments. Even when things aren't going as planned, you can help your patients to hope for simple things, like better lab results, an expanded diet order, or a better scan. And when the options for treatment run out, hope need not. Our hopes can change. From hoping for cures and treatment to hoping for more comfort, an end of suffering, or a chance to return home."

Nate's story wasn't quite over yet. A few weeks later, he came back to the hospital with a fever. His artery had clotted again. Four months after his first liver transplant, he underwent a second one. It was a deceased-donor liver from a healthy young donor who had died unexpectedly. Despite everything he had been through, Nate flew through this one. On the first postoperative day, he looked better than I had ever seen him. His hell is over now—until next year, when he starts his surgery residency.

13

Michaela

We Are All the Same on the Inside

The thing to do, it seems to me, is to prepare yourself so you can be a rainbow in somebody else's cloud. Somebody who may not look like you. May not call God the same name you call God—if they call God at all. I may not dance your dances or speak your language. But be a blessing to somebody.

—MAYA ANGELOU

What lies behind you and what lies in front of you, pales in comparison to what lies inside of you.

—RALPH WALDO EMERSON

The thing about accidents is, you don't know they're coming. Still, Lori sensed that her twenty-six-year-old son wasn't going to be around long, even though he was turning his life around. Every night, C.L. would lie on her bed and they would talk. He had told her he needed to get out of Rockford, get away from the life he'd been living. He had posted the same sentiment on Facebook a few weeks before. He told his mom that no matter what happened, he would always be with her. That night, it was as if she could hear his thoughts, and she felt compelled to ask him how he wanted to be buried.

C.L. looked at her, thinking about it for a minute. He didn't protest or ask why she would bring something like that up. "Why don't you cremate me?" he said. It was information she needed, and C.L. seemed to understand that.

It is hard to know exactly what happened the night of the accident. It was around 1:00 a.m., November 4. It's unclear whose fault it was, who started it, but at some point, there was a fight at a nightclub where C.L. was hanging out with his friends. Someone pulled a gun. There was shooting. C.L. and his two friends ran out to the parking lot. With the sound of gunfire piercing the night sky, they made it into their car. C.L. ended up in the backseat. At some point, he must have noticed that his friend in the passenger seat had been shot. As they careened out of the lot, the bullets kept coming. They sped off in total chaos, but the driver lost control of the wheel and the car crashed at high speed into a tree. Everything went black for C.L.

Lori spent a week in the hospital with her son. At one point she felt she could hear him saying he didn't know what to do. She saw a tear come down his left cheek. *I'm tired, Mom. I don't want to leave my kids*, she imagined him saying. "I know," Lori said. "I know you are." That whole week was a blur for Lori. She remembers sitting with C.L., praying a lot, touching his head, comforting him, trying to understand what he wanted her to do.

Surgical Interest Group, University of Wisconsin, Fall

I was sitting at the front of the auditorium, looking at my phone while vaguely listening to various speakers talk to medical students about the donation process, when I heard a young, fresh voice coming from the dais.

"My name is Michaela, and I got a liver transplant."

I looked up, startled to see this young, beautiful blond girl—she must have been nineteen or twenty—talking about how her life had been saved by someone she'd never met. By that point in my surgical career, I thought I'd heard everything, but for some reason so many aspects of Michaela's story gave me goose bumps.

MICHAELA GREW UP in the town of Spring Green, Wisconsin, about forty-five minutes from Madison, a town with an approximate population of sixteen hundred, 97.5 percent of them white and 100 percent Packers fans. Michaela had a normal childhood. She was a dancer and a swimmer, always healthy, never missed a day of school.

One Monday night, the family had tacos. The next day, Michaela felt sick and started throwing up. Everyone figured it was the tacos, and blamed Michaela's mother, Wendy, who'd made them. But Michaela kept vomiting, straight through the night and into the next day. She missed school that Wednesday, breaking her perfect attendance streak. When she was still sick on Thursday, the family knew something was truly wrong. When Michaela's dad, Michael, got home from work, they went to the local hospital. After doctors saw Michaela's labs, she was transferred immediately to Children's Hospital, in Madison. This was more than just bad tacos.

Michaela's parents spent the night with her at the hospital. Early the next morning, my partner Dr. D'Alessandro, with Beth, our pediatric coordinator, came to see them. The news was bad: Michaela was very sick. Her liver was failing. She likely had Wilson's disease, which at the time meant nothing to the family.

Wilson's disease is an autosomal recessive condition, which means that in order to get the disease, you need to acquire one copy of the abnormal gene from each parent. It is caused by a mutation in the gene that encodes a particular protein. It is quite rare, occurring in only about one out of every thirty thousand people. And as with most autosomal recessive diseases, patients cannot usually identify other family members who have it. The defective protein in the case of Wilson's disease is responsible for binding copper to a carrier protein to be shipped out of the liver and into the bile or bloodstream for disposal. Without this functional protein, copper builds up in the liver, and over time, this can lead to inflammation and liver damage. While Wilson's can present in a number of different ways, about 5 percent of patients show up to the hospital suddenly in liver failure, often in their teen years, and those patients will likely die without a transplant. Perhaps the only good news here is that children who present with fulminant (i.e., severe and sudden) liver failure from Wilson's disease are eligible for 1A status on the transplant list—that is, they go to the top of the list.

Michaela's mom signed the paperwork immediately. When the liver is as sick as Michaela's was, time is of the essence. Michaela was listed at 4:05 p.m. on Friday, November 11. Now she just had to wait for someone to die.

By Sunday, Michaela was obviously close to death. She was no longer responding. They moved her to an ICU, where she could be monitored closely. We urgently needed to find her a liver at this point, so Dr. D'Alessandro was contemplating accepting something more marginal, maybe an old liver, a fatty liver, one with a little fibrosis. While you hate to do that with such a young recipient, at the same time, it is a race against the clock.

ONE WEEK AFTER the accident, on Sunday, November 13, a doctor told Lori that C.L. was brain dead. His heart was still beating, but his brain was no longer functioning. The doctor asked about organ donation, whether C.L. might have wanted to give this gift of life. Lori had never talked to her son about this—she herself was signed up to be an organ donor, but she wasn't sure what he would have wanted. She looked at him and thought about what he used to say—that he would always be with her. She also remembered something else: C.L.'s stepdad, Gene, the man who'd helped raise him since he was three, needed a kidney. She had been separated from Gene for ten years, but he was still the one C.L. called "Daddy." If there was ever such a thing as a sign from above telling her what to do, Lori knew this was one.

"Yes," she said. "He wants to donate his organs." She knew C.L. would have wanted to donate a kidney to his daddy, but she also had an amazing feeling, a feeling that hit her so strongly it took her breath away. She thought to herself, *That's not the only blessing coming my way.*

Procurement for C.L., Rockford, Illinois, November 14, 8:00 p.m.

I wasn't at this particular procurement, but I can imagine how it must have gone. I suspect C.L. was referred to the OPO upon his admission—it is federal law that all potential donors be referred for consideration for organ donation when they meet triggers that suggest imminent death, including being on a ventilator with severe neurological injury. C.L. likely was referred when he first came to the hospital, on November 4. He was still receiving care with the goal of making him better, but at some point, his doctors must have come to the conclusion that he wasn't going to improve. He underwent brain-death testing, which includes both a physical exam and brain imaging, and it was determined that he was brain dead.

Lori's two options now were to disconnect the ventilator or to pursue organ donation. She chose the latter.

This decision undoubtedly started a flurry of activity. Numerous blood tests had to be sent, including those that analyze organ function, those that look for any infections in the blood, and those that identify blood group and more specific genetic typing that can be used to assess recipient compatibility. Various tests may have been run to assess specific organ function, including echocardiograms to look at the heart, and blood gases to assess the lungs. In C.L.'s case, his heart and lungs were not considered for transplant, due to the injury he'd suffered in the car accident. Because of Lori's inquiry about directing a kidney to C.L.'s stepfather, the coordinators contacted the OPO where Gene was listed, starting the process of confirming his blood type and eligibility, and had him brought to the hospital. A match run was conducted for each potential organ, and then phone calls were made to the appropriate surgeons to discuss their interest in C.L.'s parts.

Dr. D'Alessandro received the first call. I am sure he jumped at the offer.

My partner Jon Odorico likely was called about C.L.'s pancreas.

Jon would go through the match run and see if he could get C.L.'s pancreas with a kidney.

A different coordinator was charged with calling all the potential recipients to alert them to the offers, bring them into the hospital, let the OR know about the impending transplants, and let our fellows know that they were about to be busy for the next twenty hours.

At the same time, our techs from the tissue-typing lab were running cross matches (i.e., mixing potential recipient serum with donor blood cells)—once his mother agreed to the donation, C.L.'s blood had been shipped to our center from Rockford—to make sure the recipients wouldn't reject the organs acutely.

Once the recipients were chosen and called in, the coordinators would set up transportation for the procurement teams. This could involve teams from multiple programs and states, and usually involved multiple planes coming from different airports. The timing needs to be coordinated with the operating rooms at each hospital, which may involve calling anesthesia teams, scrub techs, and circulating nurses. Weather also has to be factored in, although, based on my experience, these pilots appear to be willing to fly in anything.

Once everything was in place, C.L. was brought to the OR. He was placed on the table and positioned, prepped, and draped like the thousands of other people across the country who underwent surgery on that November evening.

The nurses at Children's Hospital would have started going through their checklist to prepare Michaela for surgery. Dr. D'Alessandro would wait to call her down to the OR when he heard that the donor liver looked good and that the procurement team was getting ready to take it out.

Just after 8:00 p.m., they made their incision: a long midline going from the sternal notch (bottom of the neck) to the pubis. They opened C.L.'s belly and then used a sternal saw to open his chest. Retractors were placed exposing all his organs. His liver, which came into view right after the belly was opened, looked perfect. Why wouldn't it? It was programmed to live many more years, even though its host wasn't. His aorta was exposed, and a first glimpse of his kidneys and pancreas confirmed that they looked perfect as well. Everything was alive and working in perfect unison except his brain.

The team then spent the next several minutes dissecting out the vessels leading into the liver, figuring out if there would be variant anatomy. (As is the case for all organs, there is what we call "standard anatomy," which the majority of patients have, and then normal variations, which do not indicate any abnormality or risk for disease. In the liver, just over half of people have standard anatomy of the hepatic artery. The rest have something different. The most common variations include replaced arteries to the liver, meaning that either the right artery comes off a different spot in the arterial tree than normal, and travels deeper in the body, called a replaced right hepatic artery, or the left artery similarly comes off a different spot and travels more on the left side of the body, called a replaced left hepatic artery. This doesn't affect normal function of an organ, but matters a lot in transplant.) It's crucial not to injure these vessels. They need to be preserved with the organ, and plugged into the recipient when doing the transplant. In C.L.'s case, he had standard anatomy. The vessels were then looped, everything was dissected out, and cannulas were placed in the aorta and portal vein. A cross-clamp was placed on the aorta just as it came out of the heart, and anesthesia turned off their monitoring machines. The cold flush began flowing into the cannulas, and the right atrium of the heart was cut to let all C.L.'s blood be drawn up into suction devices. His blood was rapidly replaced with clear, sweetly smelling preservation solution, and his organs became pale and cold. The heart sputtered, took its last few, rather uncoordinated beats, and then went silent. It would not beat again. Ice was poured into the abdomen and chest. C.L. was no more. Now he was broken down into his parts: a liver, two kidneys, and a pancreas. And these parts were perfect.

Liver Transplant for Michaela, Madison, Wisconsin, November 15, 1:45 a.m.

Michaela was wheeled into the operating room early Tuesday morning. Dr. D'Alessandro began opening her belly just before 3:00 a.m. When he entered, he was greeted with one liter of ascites, just enough to bathe her organs in the sweat of the liver. Her liver was shrunken and cirrhotic, a sign that her disease had been going on for some time. Michaela had no inkling of the battle her liver had been waging for years, fighting against the copper that was killing her off cell by cell. Even though our organs are part of us, we don't have to understand what they do every day, what they might be going through. Each of our organs works for us in perfect concert, never missing a day and rarely complaining. We are unaware of their struggles, and generally can't feel them unless they push up against our abdominal wall and inflame the nerves there, or if they become extremely swollen.

Michaela's liver was done for. It was through with making her clotting factors and cholesterol, breaking down ammonia and other toxins. It was done with manufacturing her bile and collecting it in her gall bladder to be concentrated and then squeezed out into her intestine to help her break down fats and other foods.

Dr. D'Alessandro quickly removed Michaela's old, shrunken liver and dropped it in a basin. He asked for the glistening new liver and placed it in Michaela's belly. He proceeded to sew it into place, connecting the hepatic veins, the portal vein, the artery, the bile duct—and then it was done. It was Michaela's now. It whirred back to life, not minding at all that a new heart was pumping blood through it. It immediately started pumping bile into Michaela's intestine—not her gall bladder; we always remove this at the end of a liver transplant—and filtering and detoxifying her blood. Is it possible that the other organs in her body (her duodenum and jejunum, her kidneys and heart, and even her command center, the brain) paused for a second, wondering who the new recruit was? Perhaps the kidneys, which sometimes sputter after a transplant, offered mild resistance to their new teammate. Ultimately, though, they all accepted the new guy and moved on. These organs that grew up together in the small, white town of Spring Green weren't bothered in the least that a liver from a young black man from Rockford who had led a troubled life, even spent time in jail, had joined them. Black or white, yellow or brown, gay or straight, genius or moron, rich or poor, American or foreign—the organs look the same and will all function the same.

After just a few hours, Dr. D'Alessandro was already closing Michaela back up, hiding the organs in their container, their host, their home that was new to one of them.

One of Michaela's only memories of her hospitalization is of waking up from surgery with an incredibly strong craving for a hamburger. She found this odd, since she didn't like hamburgers, or meat in general. When she was finally able to eat one, she found it delicious.

Michaela went back to school right after Christmas break—all told, she missed about a month—and yet she was changed forever. The thought that someone she didn't know had saved her life after death was so powerful to her. She couldn't wait to meet her donor's family. So, on the day she got home from the hospital, she wrote the first of many letters.

"I just kept it really simple. It was . . . just saying who I am and what I did . . . and you saved my life, otherwise I wouldn't be here, and I thanked him." She included her senior photo with the letter. Michaela wrote four letters in all, and every day she went to the mailbox looking for a response, but for months, none came. She wanted to know something about the person who had saved her, whose liver sat inside her body. All she knew was that he was young, had spent some time in prison, and had died.

"I knew about the jail thing, but I never knew [why]," she said. "I know it's not that he killed someone, because he wouldn't have been able to get out . . . but even if he did . . . I'm not going to judge him, because he saved my life."

FOR SIX MONTHS, Lori grieved the loss of her son. She took some solace in watching her ex-husband Gene's life turn around after his kidney transplant, but she missed those nights when C.L. would plop down on her bed to talk. She felt sad that she hadn't been able to straighten him out, and sad that his children would grow up without a father. She thought about responding to Michaela's letters a few times, but something held her back. She wanted to know who was carrying around part of her son, but at the same time, she was sensitive to the possibility that the recipient wouldn't want to know about C.L., what it was he had done, what he had been like. Many emotions were running around in her head, holding her back.

Finally, six months after C.L.'s death, a coordinator from the OPO called her to check in. Did she want to reach out to one of C.L.'s recipients?

It was time, Lori thought. She wasn't a fan of writing letters, but she forced herself to do it. She kept it short, but felt compelled to write that C.L. loved to eat hamburgers.

The day the letter finally arrived, and Michaela held it in her hand, her mother, Wendy, quickly pulled out her phone and filmed her daughter reading it. Michaela's hands were shaking as she opened the envelope. She was always a bit shaky—the antirejection medicines tend to cause that—but much more so now than normally.

Michaela squealed as she yelled out, "He likes hamburgers! I knew it!" Tears streamed down her face.

The brief letter included C.L.'s name. Michaela fired up her computer and opened Google's browser. (Wouldn't you?) She typed in C.L.'s full name. At first, she saw a bunch of unrelated posts. Then she typed in his city and the word *obituary*—and there he was, staring her in the face. He looked young, serious, and . . . black. She felt startled, and wasn't sure why. She had nothing against black people; she really didn't know any. She had almost no black classmates and lived in a town that was uniformly white. She tried a few new searches and quickly found a few articles about C.L.'s death. Her shaking intensified.

There was a picture of a smashed-up car, a story about a shooting at a shady nightclub, a car chase, and more shooting. Police thought maybe the shooting was gang-related. Her heart was beating wildly. Would the gang come after her now, to finish the job? Would they come for C.L.'s liver?

After a few minutes, Michaela calmed down. She searched a bit more and found a Facebook post from C.L. dated just a week before his death. He wrote that he was done with this "Rockford shit." He wanted to grow up and be a better man. Michaela's fear turned to sorrow, and then gratitude. Then she noticed something funny. That night at the nightclub, when he fled, he was in an Equinox, which was strange. She drove an Equinox.

Michaela's first phone call with Lori was that night. The second Lori heard her voice, she knew who it was. They both started crying. They agreed that they would meet in person.

THEIR FIRST MEETING was in a Rockford restaurant. Michaela brought her parents and her boyfriend. Lori brought her whole family, including C.L.'s stepdad, Gene. Everyone was nervous at first, except Lori. She immediately ran up and gave Michaela a big hug. Lori hung on to her and yelled out, "This is my daughter! This is my daughter!" Their connection was immediate. Lori shared stories about C.L., Michaela told Lori and her family all about her life, and Gene told stories about his.

Since that time, they have gotten together on numerous occasions. Michaela keeps Lori updated on everything in her life. At one point, when Michaela had a minor issue with her liver—a little bump in her labs that ended up being nothing—Lori called multiple times a day to find out what was happening.

About three years after C.L.'s fatal accident, Lori posted on Facebook that she would always miss her boy C.L., but she would stop mourning now. She had Michaela to go out there and make the world a better place, and she would do this for C.L. Lori now knew what C.L. had meant when he said he would always be with her. Michaela was her daughter now.

Michaela's life has changed in so many ways since her transplant.

She has a real purpose. She owes that to C.L., but she doesn't see it as a burden. She has spent the last few years telling her story wherever people will listen—at schools, hospital functions, community events. Her goal is to get people to sign up to be organ donors. She posts pictures with Lori and her family, and even pictures of C.L., listing them all as family. A few people have questioned what it feels like to get a liver from a black man, a liver from a "bad guy," from someone who was in prison. But Michaela has used these questions as teachable moments: we are all the same on the inside.

Part V

The Donors

Show me a hero and I'll write you a tragedy.

—F. SCOTT FITZGERALD, *NOTEBOOK E* (1945)

Don't think of organ donation as giving up part of yourself to keep a total stranger alive. It's really a total stranger giving up almost all of themselves to keep part of you alive.

—AUTHOR UNKNOWN

| 14 |

As They Lay Dying

Donation gives a nobility to the final moments. It is not just
a pulling of the plug, a removal of machines. It is an act of
extending the gift of life, a giving back, a passing on. It is a way
to affirm life, to shout a note of victory into the face of death . . .
We are all given the gift of life, and to share this precious, this
fragile gift, is truly one of the ennobling acts of life.

—REV. EDWARD MCRAE, IN MEMORY OF HIS SON, STUART MCRAE

In order to do a transplant, before anything else, we need to have a
donor, living or recently deceased. There may come a day when we
can grow an organ in a dish, print one, make one mechanically, or
take one from a pig, but until that day comes, we continue to rely on
the altruism of our donors and their families. For the last ten years,
I have done hundreds of transplants, with many successes and a few
poignant failures, and I have been amazed at the strength of our
patients and their families. Yet no one has impressed me as much
as our donors.

There are two types of deceased donors. The most common is the brain-dead donor. These patients may have suffered a heart attack, stroke, asthma attack, accident, or trauma that led to loss of blood flow to the brain, causing it to swell. The cause of brain death is often hypoxia (loss of oxygen to brain tissues from shock, cardiac arrest, or low blood pressure from bleeding). The brain is encased in a hard shell (the skull), and if it swells to the point where it can no longer fit into that shell, it herniates out (exits the skull). In some cases, the brain may swell enough to block blood flow to it without herniation. Either way, the cells of the brain die, and the patient can be diagnosed with brain death. In brain-dead patients who are kept on a respirator, the heart continues to beat, the kidneys continue to make urine, the liver continues to make bile, but the patient is legally dead. Because of that, we can remove his organs, including the heart, in a controlled fashion, waiting until the very last second to cross-clamp the aorta and stop the heart. In this scenario, the organs remain well perfused until we can get them ready, which involves flushing the blood out and pouring ice on them to reduce their metabolic demand, in essence putting them to sleep until we are ready to plug them into their new owners. Yet only a small percentage of potential donors become brain dead.

The second type of deceased donor includes patients who have suffered the same heart attack, stroke, asthma attack, accident, trauma, or what have you, and have reached a point where they would not survive in a way that would be compatible with their wishes—perhaps they have massive brain damage, an incredibly dysfunctional heart, lungs that can't support oxygenation. The family (or, in rare cases, the patient) makes the decision to withdraw support by the removal of the ventilator and the shutting off of the pressors (the medications supporting blood pressure). Only if the patient dies quickly are we able to procure his organs for transplant. At our program, we will wait thirty minutes for the patient's lungs, liver, and pancreas, and up to two hours for his kidneys. We typically will not use the heart from these types of donors, as we think that waiting for the heart to stop beating prior to its removal will irreparably damage the organ. For those who don't die within the allotted time, our procurement team will fly back home empty-handed, and the patient will be moved back to the ICU, where he or she will go on to die over the next day or two. We call this process of organ donation DCD, or "donation after circulatory death" (as opposed to DBD, or "donation after brain death").

If DCD patients are going to die anyway, why don't we just take their organs out while they are intubated, under anesthesia? This is a complicated question. These patients, before support is withdrawn, are still alive by any definition of the term. They may be terminally ill, about to die, or irreversibly injured, but they are still alive. Typically, their families will still be with them and will want to stay until their loved ones are declared dead. (With brain-dead donors, the families have usually left.) Families of DCD donors will even be allowed into the OR to stand by the donor's head, even though the donor is prepped and draped in a sterile, cold room. The second these donors are declared dead, the families are whisked out of the room, and the transplant team rushes in to procure the organs. (After all, time is of the essence at this point.) If we were to remove all their organs, particularly their heart, before they reached circulatory death, then the cause of death would be "organ donation." Contrast this with the brain-dead patients, who are legally dead even though their hearts are still beating.

When I think about my experiences going on procurements, a few stories stand out. There was the little boy with a rare throat infection. He was two years old, just about the same age as my older daughter at the time. He had been entirely healthy, with a brother, a loving family, and all the dreams and imagination and joy little boys (and girls) experience. He'd gotten sick a few days before. At first, it didn't seem like a big deal, just a sore throat. But by the time he started wheezing, drooling, and struggling to breathe, and his parents rushed him to the hospital, it was too late. His little throat had swollen shut, and no matter how he tried, he couldn't get enough air down that narrow tube to fill his lungs with oxygen. His brain screamed out for more. His little heart tried and tried, but it just couldn't keep pumping—the heart depends on that air, and the oxygen it provides.

At the hospital, they snaked a small plastic tube down the boy's narrow airway, but his heart was barely working. They got it going again, but his brain had had enough. It was swollen and injured; it couldn't come back. Hooked up to a machine with breathing tubes and feeding tubes and IVs, the boy was technically alive, just not in the way his parents would have wanted him to live. But at least he could save some others.

Because he was not brain dead, he would be a DCD donor. His family came into the OR with us, to be with him until the last second. (Back then, we would stand in the OR with the family while life support was withdrawn, trying to blend into the walls while the family said their last good-byes. Nowadays, we wait outside in the hall or in an adjoining OR.) His parents played his bedtime music and read him his favorite bedtime story. His stuffed animals were all there as well, tucked into the crib his family had brought him in on.

The boy's doctor gently removed the breathing tube from his swollen throat, letting his mom and dad hold him close and kiss his face and cheeks as he sputtered and gasped his last couple of breaths. Then they laid him down on the table, kissed him again, and walked out—and we rapidly cut him from stem to stern and removed his beautiful organs. Initially, we were holding back tears (or not holding them back). The second we started his operation, this little boy became our patient, our donor. We had to get the organs out perfectly. We owed that to him, his family, and our recipients. Everyone involved went home and hugged their children a little tighter that night.

I still remember my first procurement as a fellow. We flew up north to get some organs. The donor was a man in his sixties who had died of a heart attack and was brain dead. When we got up there, part of our team went to the OR to get things set up, and Mike (our physician's assistant and procurement team leader) and I went to the ICU. I thought we were just going to talk to the nurses there and maybe examine the donor, but then Mike told me we would be speaking with the donor's family.

I felt my stomach go into my throat. Imagine a family sitting together by the bedside of a loved one in the ICU, mourning, and then having these vultures come in and tell them they have to take away their dad, brother, son, friend, to harvest his parts to be sent off for use by people they don't even know. (As I mentioned earlier, we don't use the word *harvest* in organ transplantation anymore; we prefer the more benign, less vulture-like *procure*.)

That day, we walked into the ICU waiting room to find about a dozen family members and friends of the donor. Despite experiencing perhaps the worst tragedy of their lives, donor families often get so much from being able to donate their loved one's organs. That day in the waiting room, some were crying, some laughing, some holding hands. When they saw us come in, their expressions lit up.

Mike started out by thanking them for their gift, and then told them how the process would work. He highlighted how many lives their loved one would be saving, and how that would happen that night and the next morning. They hung on to every word. When he finished, they asked so many questions—who were the recipients, where were they from, how long would their surgeries take, could they meet them, would the organs start working right away? It was a beautiful encounter. It wouldn't bring their loved one back, but his legacy, all the lives he would save, would make his death mean something. We asked the family about the donor, what he liked to do, his hobbies, and how he might want to be remembered. Despite the fact that we were about to cut him up and parcel him out to others, I couldn't imagine a more respectful way to honor this person. At the end, we all hugged. Then they said their good-byes to the donor, and we wheeled him out of the ICU.

Before we start a procurement we traditionally pause and say something about the donor. It usually involves the OR team, often the nurses from the ICU who have been caring for the patient, the anesthesiologists, and the surgeons. We will often read a passage or a poem, or express thoughts that may have been provided by the family. It is usually silent after that. Many of us will have tears in our eyes, but we'll also feel energized, elated even. We are the stewards of this donor's organs, the ones tasked with helping him or her make these supreme gifts possible. It is a heavy responsibility, but one we take on with the utmost respect and pride. In every other area of medicine, we spend our lives trying to fight off death, defend our patients from the ravages of disease, alleviate suffering brought on by cancer and heart attacks and trauma. Transplant is different. In this field, we take from death. Death is our starting point.

When people think about deceased organ donors—we also don't use the ghoulish, clinical term *cadaveric*—they probably think about those who die in motorcycle or car accidents or young people who suddenly have a brain bleed out of the blue. But so many donors die from medical conditions, such as the kid who dies from an allergic reaction to a bee sting or a peanut allergy. They die in situations that never should have happened—the baby donor who dies when his dad rolls over on him while they're sleeping; the young man who takes one false step in his house and tumbles backward down a staircase; and seven-year-old Caleb, whose normal day turned into the worst nightmare for his family.

Caleb was the middle child in a family of five, two years younger than his hero, Cole, and two years older than his sister, Katie. He was a happy, kindhearted kid who loved to hug everyone in his family every chance he got. This particular day, a Sunday in December, started off a bit too early. The children's cousin was staying over, sleeping in the room Cole and Caleb shared, and the boys, excited to start the day of play, as most young children are, got up at the crack of dawn and came into the room where their parents, Dan and LeAnn, were sleeping. LeAnn told the boys to go back to the living room and play quietly until it was time to get up for church. She suggested they do some coloring, or whatever they wanted—as long as they didn't wake their sister. Then she drifted back to sleep—only to wake up when she heard Caleb urgently pulling at the sheets on Dan's side of the bed. He was choking on something; she couldn't quite understand what he was saying. At first he was talking, but then he couldn't make any sound.

They called 911 and brought Caleb downstairs. The ambulance took forever, probably because of the ice storm of the night before. Those minutes of waiting were excruciating for Dan and LeAnn. They kept thinking of throwing Caleb in their car and driving him to the hospital themselves, but then thought better of it. By the time the ambulance arrived, Caleb looked blue. They could tell from the look in one of the paramedics' eyes that their son was in trouble. He was brought to the local hospital and then quickly transferred to Madison, where he was taken to the OR. Doctors there found a small green tack wedged in his airway. It was small but it had wedged itself in perfectly, completely obstructing the boy's air flow. Over the next two days, Caleb remained intubated, with machines breathing for him.

At first it seemed that he might get better; he had been kept in a medically induced coma, to let his brain recover from the lack of oxygen. The team told the extended family, who had been sitting vigil at his bedside, that they would try to wake him the next day. They told everyone to go home and get a good night's sleep. After everyone else left, LeAnn went to the bathroom, and when she returned to Caleb's bedside, the alarms on his monitors started going off. Caleb's blood pressure and heart rate had shot up and then plummeted. She knew he was gone. That was the moment Caleb's brain herniated.

The entire family was brought into a conference room that night and told that Caleb was brain dead. They were devastated. When they were asked if they would consider donating his organs for transplantation, Dan and LeAnn said yes immediately. They needed something positive to come out of their son's death.

Ultimately, eight organs were transplanted from Caleb's body into recipients who were waiting for the gift of life: his heart, both lungs, his liver (which was split between a baby and an adult), both kidneys, his pancreas, and his small bowel. So many lives were saved by this one boy who had died for no good reason on a day like any other.

Much time has passed since that horrible day in December. Dan and LeAnn are doing well, spending their time with their two other children, making memories they will cherish forever. Those memories will include stories about Caleb. They won't think about the horrible times as much as the good, and maybe their memories will include stories about the recipients of Caleb's organs. We hope his lungs are filling with air every couple of seconds, sending oxygen through someone's young body. We hope Caleb's heart is out there somewhere, still beating away, pumping blood around some boy or girl, giving him or her enough strength to run around a playground. Maybe someday LeAnn and Dan will get to listen to its rhythmic sound.

AND THEN THERE was Kylie, whose mother, Shirley, introduced herself to me in this way: "We are a family of five, my husband, Bruce, and I. And we have three children: Kylie, our oldest daughter, she passed away at age seventeen due to a fatal car accident; my son, Chase, he is nineteen; and our youngest daughter, Kensie, is now seventeen." In Shirley's mind, they were still a family of five.

The day Kylie left them, a day forever preserved in their memory, was a Sunday in summer, one of those July days when the warm, gentle wind blows just enough to make it comfortable; the type of day when the screen doors allow the breeze to move through the house, beckoning you outside. Normally, on a day like that, Shirley and her family would have been out on the river. But Shirley knew Kylie would be back home at around noon, after a weekend away at a wedding with her boyfriend and his family, and she wanted to hear all about it. Kylie had earlier asked her mom if she would go with her to look at some spots around town where she could have her senior pictures taken. Shirley couldn't believe her older daughter was already making plans for college.

After church, Shirley did a little shopping, picking up some shorts Kylie had asked for. A little after noon, her son, Chase, came in and told her that a MedLink helicopter had just landed in a nearby field. He thought maybe there'd been a farm accident.

This didn't sit right with Shirley. Kylie should have been home about now. She sent her a text, but heard nothing back. So, Shirley and Bruce made their way to the scene of the accident—where their greatest fears were realized: Kylie's car was wrapped around a tree and surrounded by emergency workers.

The next few hours went by like a blur. Kylie was so unstable that she barely made it to the hospital. Shirley and Bruce sat in the waiting room, receiving periodic updates from the nurses, none of which brought any hope. Kylie wasn't even stable enough to be brought to the CT scanner.

Kylie never made it out of the trauma room. She was declared brain dead. And before they knew what to think, Shirley and Bruce were asked about organ donation. The decision was easy. "Kylie had told me when she had gotten her license that she was going to be an organ donor, and we had talked about it at that time. 'Mom,' she said, 'when the time comes, and I go, I don't need my organs for anything. So why not give them to somebody? Why not try to save a life or two lives or three lives or how many lives you can save?' So, she was actually the one that brought it up and that made me change it on my license to an organ donor."

Kylie was moved to an ICU, where family and friends could stay close to her. She was extremely unstable, and the doctors and nurses were working hard to keep her heart beating. The wait for our procurement team to come up from Madison was interminable for Shirley and the rest of the family.

"Well, I knew that it was Kylie's wish," Shirley told me, "and I knew that I wasn't going to get her back. At that point I wanted her heart to keep going long enough to fulfill her wish. So, I think if her heart would have stopped, and they weren't able to do the organ donation, it would have been even more sad for me."

The workup of a brain-dead donor might take twenty-four to thirty-six hours before we would be ready to procure the organs. In addition to placing the organs in hospitals all around the country, sometimes invasive tests are needed (such as a liver biopsy or a cardiac catheterization of the heart). This is usually done before the deployment of a procurement team, which must wait for the results of the blood work indicating donor blood type, organ function, and any infections the donor might have been exposed to, including hepatitis C or HIV. Kylie's accident took place just after noon. Organ procurement began at around 10:00 p.m. It was that quick.

Kylie's case was what we call a "drop and run." Because she was so unstable, before any of the blood work is done, we send a procurement team to get the organs. In these cases, we usually take only the kidneys—we can put them on ice in a cooler or on a pump that flushes solution into them—while we wait for the test results. Then we have the recipients brought in, and perhaps do the transplants twelve to twenty-four hours later.

When our team arrived, Shirley and Bruce went in to say goodbye to Kylie. She was already brain dead, but it would be their last chance to see their beloved, beautiful daughter while she still had a beating heart. It was all very rushed, but Shirley remembers the moment vividly, standing with her husband, gowned up in OR attire, staring at her daughter for the last time. Then they waited until the procurement was done, when Kylie—without her kidneys (which our team took) or eyes (which would be used for a cornea transplant), and with her heart no longer beating—would be returned to the room before being brought down to the morgue.

"I pretty much had guessed that they weren't going to be able to use her heart for anything," Shirley said, "since it had stopped so many times. I had kind of assumed that it would be way too damaged, and that, of course, was the case." Still, she said, "I would have loved for somebody to have gotten Kylie's heart."

Shirley and her family still think about Kylie every day. It has been a struggle, and they are just now getting back to living the way Kylie would have wanted them to. I asked Shirley if the donation had helped with her recovery.

She said it had: "I think that it's a way for Kylie to have lived on and for people to remember." She added, "And, you know, it has been a good experience, and it helps me. It helps me to know that she wanted to be a donor. She did something fantastic for somebody else and made that person's life so much better and, you know, I have a connection to this person. And through that connection, it just makes me feel like she's always still here."

FOR TRANSPLANTATION TO succeed, implant operations and immunosuppression strategies had to be defined and then revised over many decades. Yet those efforts would have been in vain without a better understanding of death and dying, of what really constitutes death. Transplantation, out of necessity, would serve as a catalyst for further defining society's feelings on life, death, and the fine line that sometimes separates them.

Louvain, Belgium, June 3, 1963

Guy P. J. Alexandre had just returned to Belgium after spending his third year of residency with Joe Murray in Boston. He came directly after Roy Calne to Murray's lab, and was the one who prepared the azathioprine solution for the successful human transplants Murray performed in 1962. When he flew back to Belgium to complete his training, Alexandre's luggage contained vials of azathioprine and other promising immunosuppressive agents. (Yes, it was a simpler time.)

Over the next year, Alexandre completed his surgical training, and then convinced his team that they should move forward with clinical kidney transplantation. "All that remained was to find a suitable patient and to maintain him in a suitable condition until a kidney could be obtained and transplanted." As he recalls:

> On June 3, 1963, a patient was brought in with a head injury and in profound coma. The patient became completely areactive, had a falling blood pressure despite the administration of vasopressive drugs, and presented all the signs of what Mollaret in 1959 had described as a 'coma depasse' (a state beyond spontaneous breathing) . . . Professor Morelle [the chair of the department of surgery], who was quite experienced in neurosurgery, considered the neurological symptoms presented by the patient and took what today could be considered the most important decision of his career: whether to remove a kidney from that patient while the heart was still beating. This procedure was probably the first transplant ever removed from a heart-beating cadaver. Fortunately, this was long before the days of established ethical committees.

Alexandre transplanted the kidney into the recipient, whom he had been operating on in the room next door. The transplanted kidney started making urine right on the table, and the patient had normal creatinine levels a few days later.

This marked a huge turning point in transplant history.

London, March 9, 1966

Transplant pioneer Michael Woodruff, chair of surgery at the University of Edinburgh, had seen fit to organize a conference to discuss ethical and legal issues in transplant. Attendees included Joe Murray, Tom Starzl, and Roy Calne, along with the majority of players in the small but burgeoning field of transplantation. Among the many topics Woodruff wanted to discuss at the conference was the problem of obtaining donors. In the early days of transplantation, meaning the 1950s and early '60s, so few transplants were being conducted that enough organs could be obtained from living donors, patients having their kidneys removed for other reasons, and patients who had just died (including prisoners). Of course, obtaining organs from these donors was no small task. Surgeons would have to wait until a donor's heart had stopped and he was declared dead (meaning no heartbeat, no blood pressure, no respiration). At that point, consent would be obtained from the family. Then the patient could be taken to the OR for procurement of the kidneys. A significant amount of time would pass before the kidneys were removed, and all the while they would get no blood flow or oxygen. Everyone knew that wasn't great for the organ.

In England, Calne was often forbidden from bringing these donors to the OR for procurement:

We knew that kidneys had to be removed from the dead as quickly as possible after the heart had stopped beating, otherwise the organs would be useless. But here we hit a serious snag: the nurse superintendent of the operating theatres would not permit dead bodies to be brought into her operating rooms. So we had to remove the kidneys in the open wards. Looking back on the procedure, it must have re-sembled a horror film. Other patients in these large wards would see a team of surgeons rush in, go behind a curtain where a patient had died, and operate on the corpse in an ordinary hospital bed. This was very difficult and blood would often trickle onto the floor, where the patients would see it and be further terrified and upset.

At the symposium, after Joe Murray delivered his lecture titled "Organ Transplantation: The Practical Possibilities," a discussion ensued. At some point, Alexandre got up to make a statement. "To throw some fuel into the discussion, I would like to tell you what we consider as death when we have potential donors who have severe craniocerebral injuries. In nine cases we have used patients with head injuries, whose hearts had not stopped, to do kidney trans-plantations." He went on to describe the very specific conditions these donors needed to meet, including the absence of reflexes, the absence of response to pain, a flat EEG, and the absence of spontaneous respiration five minutes after the ventilator had been withdrawn—in other words, he gave the definition of brain death without using that term. His team simply called it death.

Alexandre's remarks were a bombshell. Starzl said, "I doubt if any of the members of our transplantation team could accept a person as being dead as long as there was a heartbeat. We have been discussing this practice in relation to renal homografts. Here, a mistake in evaluation of the 'living cadaver' might not necessar-ily lead to an avoidable death since one kidney could be left. But what if the liver or heart were removed? Would any physician be willing to remove an unpaired vital organ before circulation had stopped?"

Calne went further: "Although Dr. Alexandre's criteria are medically persuasive, according to traditional definitions of death, he is in fact removing kidneys from live donors. I feel that if a patient has a heartbeat, he cannot be regarded as a cadaver."

As Alexandre recounted years later, "at the end of the CIBA symposium meeting . . . the president of the meeting asked the participants to let him know those who were prepared to act the way we were doing and to accept our criteria of brain death; I was the only one to raise my hand, all the others did not." Still, despite his inability to convince anyone in the room that he was right, Alexandre did plant a seed in the mind of Joe Murray, perhaps the most prominent transplant surgeon in the world at the greatest institution for transplantation in that era.

Over the next few years, efforts at transplantation continued to grow, primarily with the kidney but also the liver. Then, in 1967, the first heart transplant was performed. While it remained the standard in the United States that donors had to be declared dead by a physician prior to organ procurement, the notion of what constituted death was slowly shifting. Some surgeons had taken Alexandre's lead and were removing kidneys from patients with coma dépassé, although they weren't publicizing it. As Murray recounts in his autobiography, "Across the country, the response to the seemingly simple question 'When is a person dead?' varied widely. Most alarming was the position of certain doctors at the kidney conference [not clear which conference] who had stood up and said, 'I'm not going to wait for the medical examiner to declare the patient dead; I'm just going to take the organ.'" This concerned Murray greatly.

Surgeons had recognized by that point that donors with beating hearts were preferable in terms not only of timing and availability but also of outcome. When the donors had beating hearts, the organs continued to receive blood flow and oxygen, and continued to function, until the very moment that they were removed. This made them more likely to work immediately following transplant, and continue to work thereafter. Yet as the successful transplanting of organs (including livers and, ultimately, hearts) was becoming a reality, the importance of finding consensus on this ethical issue became ever more urgent. (It didn't help that the press had begun accusing the Brigham physicians of playing god in their efforts at organ transplantation.)

Boston, Massachusetts, September 1967–June 1968

Henry Knowles Beecher would play a central role in the definition of brain death. Beecher was a prodigy, a true genius, and his uncommon intelligence was recognized by all who knew him. He became anesthetist in chief at the MGH in 1936, and served in North Africa and Italy during World War II, an experience that affected him greatly. He was one of the first researchers to write about the placebo effect (in 1954), underlining the importance of the double-blind, placebo-controlled trials that are the gold standard today. In 1966, Beecher became both famous and infamous when he published a paper in *The New England Journal of Medicine* entitled "Ethics and Clinical Research," which highlighted multiple cases of clinical research that placed the subjects at risk of morbidity and death without true informed consent. Although controversial, this report was one of the first to lay the groundwork for the guidelines on informed consent that we use today.

In September 1967, Beecher wrote to Robert Ebert, dean of

Harvard Medical School, asking him to convene a meeting of the school's Standing Committee on Human Studies to discuss the topic "Ethical Problems Created by the Hopelessly Unconscious Patient." Beecher wrote, "As I am sure you are aware, the developments in resuscitative and supportive therapy have led to many desperate efforts to save the dying patient. Sometimes all that is rescued is a decerebrated individual. These individuals are increasing in numbers over the land and there are a number of problems which should be faced up to."

The meeting was held on October 19, 1967. One of the attendees was Joe Murray, who suggested that a new definition of death be codified, and that Harvard was the place to do it. Beecher reached out to Murray in a letter the next day, thanking him for his comments and agreeing that it should be done at Harvard. Murray wrote back a week later: "The subject has been thoroughly worked over in the past several years, and by now areas for action are crystallized into two categories. First is the dying patient, and second, distinct and unrelated, is the need for organs for transplantation." He went on to identify the necessity for "a medical definition of death" and the need to enlist the "opinions of the neurologists, neurosurgeons, anesthetists, general surgeons and physicians who deal with terminal patients."

The next paragraph of Murray's letter highlighted the challenges he was dealing with in transplantation:

The next question posed by your manuscript, namely "Can society afford to lose organs that are now being buried?" is the most important one of all. Patients are stacked up in every hospital in Boston and all over the world waiting for suitable donor kidneys. At the same time patients are being brought in dead to emergency wards and potentially useful kidneys are being discarded. This discrepancy between supply and demand is soluble without any further medical knowledge; it requires merely an educational program aimed at the medical profession, the legal profession, and the general public.

The committee had its first meeting on March 14, 1968, and worked on six drafts, completing its report on June 25, 1968. Murray played a prominent role in the writing, and he and Beecher talked almost every day during those three months. The final document was published on August 5, 1968, in *JAMA*, under the title "A Definition of Irreversible Coma—Report of the Ad Hoc Committee of the Harvard Medical School to Examine the Definition of Brain Death." The first line of the paper states, "Our primary purpose is to define irreversible coma as a new criterion for death." The authors go on to discuss the importance of this for two reasons: first, to deal with futile care of patients in the ICU, and second, "Obsolete criteria for the definition of death can lead to controversy in obtaining organs for transplantation."

As David Hamilton writes in his history of organ transplantation, "In normal circumstances, this much-needed new definition of death might have encouraged a slow but orderly change in medical practice. But the times were not normal: Barnard's transplantation of a human heart, performed just after the Harvard committee had been launched, had changed everything." Unfortunately, the outcomes of these heart transplants were truly awful in those early years, and the public became frightened by the idea of hungry heart surgeons removing their loved ones' beating hearts while they were still alive. Public support for heart transplantation, which was at an all-time high when Christiaan Barnard performed the first in December 1967, was at a low by May 1968. Numerous hospitals stopped allowing organ donation, and acceptance of the concept of brain death in the United States was definitely not immediate. It took a little more than a decade for public opinion to align with that of the Harvard committee and accept the definition of brain death.

THE PIONEERS IN transplant were willing to persist even with not only the very clear reality that their colleagues thought they were crazy, and outright killing people, but also the knowledge that they could end up in jail. Why were they willing to risk jail time to make transplant a reality?

The courage and commitment this required are palpable for me when I think of my patient Wayne, whom I first met when he had just been diagnosed with ALS (amyotrophic lateral sclerosis, or Lou Gehrig's disease). He found himself getting progressively weaker, and eventually ended up in a wheelchair. When he could no longer hold his head up or swallow his own saliva, he knew death was imminent.

I had met Wayne a couple of years before this, when he was considering donating a kidney for transplant; he wanted to make his disease mean something for someone. We came up with a plan: we would bring him to the hospital when he got to the point where the quality of his life was so bad that he knew he couldn't go on. We'd admit him to my service, bring him to the operating room, and remove one of his kidneys (or possibly both) for donation. Then he would be returned to the ICU, and once his anesthesia wore off, we would extubate him and let him die naturally of ALS. This way we wouldn't be killing him for organ donation, but he would be able to give the gift of donation that was so important to him.

As Wayne deteriorated, we convened an ethics meeting at our hospital to discuss the situation. The meeting was well attended, and support for proceeding was overwhelmingly strong. Then, as we were putting the details in place, I received an analysis by our lawyers of the risks of proceeding. They stated that there was a significant likelihood, if we went ahead with the plan, that we (I) would be charged with murder, or at least acceleration of death, a charge that would be left to the discretion of the district attorney for our state. As I read this report and thought about it, I realized there was no way in hell I would proceed—not that the hospital would have let me anyway, given the legal department's response. Thoughts of getting charged, going through a trial, and living with that risk were all too much for me to bear. I wasn't going to jail for this. I had kids to think about!

When I told this story to Tom Starzl, he had a short response: "Well, you just explained exactly how not to go forward . . . We just did it," he added, referring to the transplant pioneers. "These were sneak attacks that took place. They were done before the naysayers had a chance to say nay." We all owe a debt of gratitude to Starzl and others like him for their courage.

In 1980, the Uniform Determination of Death Act passed in the legislatures of all fifty U.S. states, declaring brain death legally equivalent to death. This legislation has been critical to the success of organ transplantation. Brain-dead donors give us the most organs, enable the best outcomes, and offer surgeons the most controlled scenario to perform organ transplantation. Brain death's being equivalent to death has been the law of the land here in the United States, and in most countries, ever since, and in general the public has accepted and agreed with that.

| 15 |

Healthy Donors

Do No Harm

Practice two things in your dealings with disease: either help or do not harm the patient.

—*EPIDEMICS*, BOOK 1, HIPPOCRATIC SCHOOL

The ordinary man is involved in action, the hero acts. An immense difference.

—HENRY MILLER

I think by and large a third of people are villains, a third are cowards, and a third are heroes. Now, a villain and a coward can choose to be a hero, but they've got to make that choice.

—TOM HANKS

There is no other discipline in health care in which we perform an operation on someone who has no medical diagnosis, no identifiable pathology, who is by definition entirely healthy and will get no medical benefit and even some clear harm from the procedure. There is even a chance these patients could die, although this is rare (about three in ten thousand for kidney donors). Yet caring for these particular patients is where I get the most satisfaction, feel the most pride, and lose by far the most sleep.

While the risk of death is low, other things can happen. We can nick their bowels (I have), rupture their spleens (I have), get into the bladder (done that), get into bleeding (of course I have), have to convert rapidly to an open surgery (yup), and make a hole in their diaphragm (sadly, yes). All these mistakes are fixable (and I fixed them all).

After surgery, the donor can develop high blood pressure because of his donation, and even down the road can go into renal failure (about a 1 percent chance, which is lower than the general population but higher than if they hadn't donated). So, how do donors feel about these risks, and the idea of donating in the first place? Most of them are grateful to have the opportunity.

The National Kidney Registry (NKR) is an organization that facilitates paired kidney exchange. When patients in need of transplants have living donors who can't give to them directly (because of blood type incompatibility or antibodies present in the recipient blood), the NKR facilitates swaps. Often this involves more than two pairs—there can be three, four, or even more. This complex exchange is aided by computer algorithms. These donors and recipients may reside in different cities, sometimes across the country. In these scenarios, I may go to the operating room in Madison at 6:00 a.m., take a kidney out, and send it on a plane to New York. Conversely, someone in California may take one out in the evening and send it on its way to the East Coast on a red-eye. In a second scenario, when a humanitarian decides to donate a kidney to help a stranger in need, this can start a chain. One woman's kidney can go to a recipient whose incompatible donor then donates to someone else, whose donor donates to someone else, and so on. The chain can go back and forth, crisscrossing the nation for weeks on end, until it finally breaks. The longest chain the NKR has had to date included thirty-four donors and thirty-four recipients, involved multiple transplant centers across the country, lasted approximately three months, and (interesting for us) ended at our center. It started with one person who wanted to give the gift of life. How amazing is that?

At an NKR fund-raising gala in New York City a few years ago, a young woman got up and took to the microphone. She seemed a bit nervous at first, but then delivered a moving, unforgettable speech: "To all you doctors out there, you guys get to save people's lives every day. But for me, I don't get to do things like that. I have a pretty normal life. But one day last year, I donated a kidney, which started a chain that saved more than twenty people. Sure, I was in the hospital a few days, and felt sore for a couple of weeks. But I can honestly say it was and will always be the best thing I have ever done in my life. I think about it every day." This young woman was truly heroic.

When I see potential donors in my clinic, once I have gone through their medical history, examined them, and looked over their labs, I typically have a conversation with them about the risks. I tell them that most patients do great—almost all, actually—but that there is always the risk of this or that. And most patients say, "Sure, I get it. It's surgery."

I struggle with using percentages. If I tell a patient he has a 1 percent, even a 5 percent, chance of dying on the table, most patients will say, "Great, that's not going to happen to me." But for me, that number is high. It means that some of my patients absolutely will die on the table. Of course, this is higher than the risk our donors face but is closer to the risk of liver transplant recipients.

Nancy Ascher, chairman of surgery at the University of California–San Francisco and one of the premier transplant surgeons in the country, gave a kidney to her sister a few years back. A couple of days after the surgery, Dr. Ascher had to be returned to the OR emergently: her bowel had gotten caught in one of her incisions and become obstructed. Uncommon, but pity the poor surgeon who operated on his chairman!

When Dr. Ascher talks about her donation, despite knowing all the data better than anyone, she still describes the experience as a "leap of faith." Regardless of the statistics, you are putting your life in someone else's hands. If something goes wrong—if a harmonic scalpel strays too far one way or another—your life can be changed forever. Dr. Ascher does not regret her decision. On the contrary, she is proud she did it. She now believes in the concept of living donation more than ever, but her experience changed how she talks to patients, and how she thinks about risks and recovery.

Here's an example of why the numbers don't always affect a donor's decision. A potential donor came to see me wanting to donate a kidney to his wife, the love of his life, the mother of his children. He had been turned down by a major program due to medical reasons. He didn't have any absolute contraindications—he was not diabetic, didn't have heart disease, didn't have any cancers—but he was obese, his blood pressure was high, and he had a history of smoking. Each one of those things could have increased his risk of having a problem with his remaining kidney someday.

When I first saw this man, I had my doubts, too. But he told me,

·

"I know you guys think I'm a little more high-risk than your average donor. I get that. But you've gotta help me. This is something I really want to do. My wife is my whole life. Our family doesn't work without her. She has done so much for me. If you let me do this, I will thank you forever. I'll sign any form you want. I promise I won't sue you. If, down the road, something happens to my other kidney, and I go on dialysis, it will be worth it for me. I have no hesitation."

What do you say to that? This man seemed to have grasped the data, understood the risks I'd outlined for him. Wasn't it *his* body? How paternalistic was I supposed to be? Of course, it was possible for someone to make a bad, even an unreasonable decision for himself to protect someone he loved. Was there a risk that was so high that we would have to say no? We wouldn't let mothers and fathers donate their hearts to their children, even though many patients would be willing to do this.

In the end, I let the guy donate, but not before giving him my standard line. It's not necessarily data-based, but I believe it: "If you donate and stop smoking, you may actually be healthier than if you don't donate and keep smoking." (Of course, the worst scenario was the most likely one: that he'd donate and keep smoking.)

His surgery was difficult—he was quite big—and he had some tough days during his recovery. But ultimately, everything went fine (for now). (His wife did great, too.) Back in my clinic about a month after the surgery, I asked him how he felt about the whole process. "Doc, this is the best thing I have ever done in my life. Sure I had a few bad days, but I have my wife back. I will never be able to thank you enough for what you did for my family." He gave me a big, warm handshake. His thick, ruddy hand enveloped mine. It was the hand of someone who had done physical labor all his life, a hand that would provide for his family no matter how much it took from him.

He got one thing wrong. I'm not the one who did something for his family.

REMOVAL OF A donor's kidney is done laparoscopically, so it's a bit different from the other procedures I have described. Once the patient is asleep on the OR table, she—women donate far more than men; yet another example of how women are the superior sex—is turned on her side. If we are taking the left kidney, we place the patient on her right side, and vice versa. More often than not, we take the left kidney, because the vein there is longer and easier to implant. Then we carefully poke a needle in the left lower quadrant, lateral to the belly button (but a little lower), feeling a few pops as the needle gently enters the abdomen. We then insufflate CO_2 and blow up the belly with gas. This allows us to have some space to see.

At this point, we insert our first working port, with a camera inside it so we can monitor our progress as we slowly advance the port, watching the tissue planes, going through fascia and muscle, until we enter the abdomen safely. After this port is in, we put in our other working ports. We then insert long, thin instruments through the ports, again watching our efforts with the five-millimeter camera in the first port. Typically, the surgeon uses two instruments: in the left hand, a grasper or a suction; and in the right, a harmonic scalpel. This instrument with the cool-sounding name has two blades at the end that vibrate ridiculously fast (fifty-five thousand vibrations per second) when you press a button. We use this instrument to dissect (your fingers go in the handles like a pair of scissors), and to divide structures, including blood vessels; the vibrations seal small vessels.

In the first part of the operation, we mobilize the colon out of the way, to get to the kidney and ureter, which sit behind it. This part is usually easy. Then we mobilize the spleen and its attachments away from the kidney. You have to be very careful not to nick or injure the spleen—this organ is basically a bag of blood, and the smallest injury can bleed enough to require its removal. Once the kidney is exposed, you have to dissect out the blood vessels. Normally, there is one vein and one artery, but there can be variations on this. We always know the anatomy beforehand, from the patient's CT scans, which delineate these structures for us. Sometimes kidneys can have two or three arteries and veins. Dissecting these vessels is both fun and mildly stressful. You advance slowly, spreading tissue by opening and closing your hand as you watch the tip of the instrument on the TV screen in front of you. It's critical not to make a hole in one of the vessels, as the bleeding can be tricky to control and makes it hard to see anything with the camera. First, we dissect out the vein, and then the artery. Then we peel the adrenal gland off the kidney and divide the adrenal vein. Now we go to the back side of the kidney, and free it up from its attachments posteriorly. We then flip it over and make sure it is only attached to its vessels and the ureter.

Okay, now we are ready to make a small incision for the organ's removal. We make a small cut below the patient's pant line, maybe just three or four inches in length horizontally—the same incision used for C-sections. We then open the peritoneum and insert a metal tube that houses an expandable bag. Then we put our camera back in and get ready to remove the bean.

First, we make sure they are ready next door in the implantation room, with ice, and UW solution hanging to flush the kidney out. Then we insert a linear stapler through the belly button port. This stapler has big jaws, which you put around the ureter, artery, and vein separately. When you push its button, it fires a row of titanium staples—actually three rows on the "stay" side and three on the "go" side ("stay" side is the side that stays in the body, "go" side comes out with the kidney)—and cuts between them. After each fire, we remove the stapler and hand it back to the scrub tech, who reloads it and hands it back. Once all the vessels have been disconnected, we push the bag out of its metal tube into the belly—while we watch it on the camera—causing the bag to expand so we can scoop up the kidney like a goldfish. We pull the bag's drawstring, closing it. Then we pull the bag with the fish—er, I mean kidney—out of the belly and hand it off to the recipient surgeon, who will cut off the staple lines, flush the blood out with cold UW solution, and get the kidney ready for implantation.

Still in the donor room, we will look for any bleeding, close up our ports, and get out of Dodge. Easy-peasy—sometimes. That's the thing about surgery: some cases are easy; some aren't. Some people are thin on the inside, some are filled with internal fat, making all the planes tricky. When I do a laparoscopy, I study the films carefully beforehand, developing a 3-D image of the patient's anatomy in my head. As I dissect, watching on a TV screen in 2-D, I picture where everything is, what is behind what. If some structure doesn't look right or isn't where I expected it to be, or if I lose my mental picture and can't predict what I am about to see, I slow down, take baby steps, millimeter by millimeter, until it makes sense to me again.

The vast majority of these cases go well, but there is always a chance of injury. We surgeons spend our lives learning how to get out of problems when we get into them. All of us can handle the easy cases, but it's the expert surgeon who can make a hard case look easy, avoid problems others get into, and most important, get out of the problems that inevitably occur. Another mark of the expert surgeon is knowing when to call for help. Pride has no place in the operating room.

So those are the steps of the donor operation, a surgery I have performed hundreds of times. In addition, I have evaluated close to a thousand people interested in donating kidneys. In some ways these surgeries and evaluations have become almost commonplace to me, as they are reliably on my schedule every week. And yet every time I meet a potential donor or begin a donor operation, I find myself amazed by the altruism and bravery of all these heroes. I am not just awed that they are donating parts of their bodies; they are actually allowing themselves to become vulnerable patients, so their recipients don't have to suffer alone. To me, one of the worst parts of being ill is how it separates you from your loved ones, leaving you isolated from everyone and everything that matters to you. When you are really sick, dying sick, you have to accept that you won't get to watch your children grow up, get jobs, get married; that you won't get to be the person you might have become. Sure, your friends and family may mourn you for a while, they may think about you from time to time, but life will march on. Living donation allows a loved one to take your hand and say, "Let's do this together." The risk the donor takes, however small, is an important part of the equation. It says, "I will be sick with you, and together, we will fight through this. I will take that same leap of faith you are taking, put my life in the hands of someone we don't know, and the two of us will be stronger than one."

Kidneys are not the only organs that can be transplanted from living donors; livers, or portions of livers, also can be transplanted this way. But there are some major differences between these two types of transplants, including the severity of illness of the recipients, the organ allocation system, and outcomes between living and deceased organs. When your kidneys fail, you go on dialysis and then (assuming you are referred for transplant) on the kidney transplant list, your place on the list determined primarily by how much time you've spent on it: the longer you wait, the closer you get to receiving a kidney. But that waiting list is for deceased-donor kidneys.

Living-donor kidneys are different, and better, for a few reasons. First, you don't have to wait on a list for them. You bring your own donor and you get transplanted, maybe even before you have to go on dialysis. Second, living-donor kidneys last longer. The half-life for a living-donor kidney is fifteen years, and many do better than that. For a deceased donor, it's about eight to ten years.

When your liver fails, there is really no dialysis. When a liver is working really badly, you get increasingly sick until you die. The only treatment is a liver transplant. Unlike kidneys, livers are given out based on how sick you are as determined by lab values, and that score has been shown to predict how close you are to death. As your score goes up, the higher up you go on the list. The hope is that a transplant takes place before you are too sick to survive. In general, living-donor livers—for these, half the donor's liver is removed and placed in the recipient—perform roughly the same as deceased-donor livers in terms of survival, although the operation to plug in a living-donor liver is certainly more complex than that for a deceased donor (from whom you get the whole liver).

For donors, the story is a bit different. For liver donors, risk of death is somewhere between 1 in 200 and 1 in 600; the risk of complication may be as high as 30 percent; and the length of stay is about a week. As long as you avoid early problems, you should live a normal life. In the United States we do more than 5,000 living-donor kidney transplants per year, and something like 250 living-donor liver transplants.

Do I have the same wonderful feelings about living donation for liver as I do for kidney? As long as the donor knows what she is getting into, the answer is yes. But I do have some hesitation. Patients tend to have a very different assessment of risk from surgeons. I look at a death rate of between 1 in 200 and 1 in 600 as high. Potential living donors need to understand that they could die donating part of their liver. They could die donating a kidney as well, but it is less likely by a factor of 10.

For surgeons, a critical part of the conversation when getting the consent of a living donor is the concept of coercion. While the vast majority of donors really do want to donate, seeing a family member suffering, even at risk of dying, can create undue pressure. Imagine a doctor telling you that you could save your family member if you were willing to take a little risk, endure a little pain, accommodate a little disruption in your life. Imagine the pressure other family members might put on you to save your sister, your father, your son. This is why all programs in the United States require the presence of an independent coordinator, who meets with potential donors to explore donor motivation and conviction. Still, in the end, family dynamics always plays a role in these decisions.

Charlie Miller, the director of liver transplantation at the Cleveland Clinic, is widely regarded as an incredibly gifted surgeon who makes the hardest cases look easy. Still, everyone in the field knows that as the head of liver transplantation at Mount Sinai in New York City back in the early 2000s, he had a donor death. A young man was donating a portion of his liver to his brother. The surgery itself went fine, but a few days later, the donor died. Numerous errors were documented in the newspapers and tabloids, citing "woefully inadequate" care. While Sinai took a big hit in the unforgiving New

York press, Miller himself got eviscerated. Amid the various investigations, he ended up stepping down from his role at Sinai and disappeared from view for a couple of years. Eventually, he turned up at the Cleveland Clinic, and revived his career and his standing—as is perhaps best exemplified by his becoming the president of our transplant society in 2015. He continues to be involved in living donations.

In one of the most meaningful talks I have heard as a transplant surgeon, Miller documented his experiences at Mount Sinai—who he was before the death, how filled with confidence he was, how sure he was that he would never have a donor die on him. Sure, he knew the data. But he was so good, he thought, that those numbers didn't apply to him. When it happened, of course he was devastated. But at the same time, it seemed like a freak thing—the donor had gotten a rare infection in his stomach, likely related to some food he ate, maybe related to the surgery. It didn't feel like a surgical death: the donor didn't bleed out on the table or lose his liver function because Miller had tied off a vessel.

Miller was unprepared for what happened next. According to a *New York Times* article that came out after he was hired at Cleveland, "Almost instantly, it seemed, he went from the top of his profession to being 'almost nothing,' he said, as if his entire career had been erased." He went to Japan for nine months, and then ended up in Modena, Italy, "at the invitation of a friend and colleague who knew that a fellow surgeon needed to be in the operating room." As Miller talked about rebuilding his life, his confidence, his career, he also described the despair he felt at the time. He was fifty-one, and faced with the prospect of losing a career that he had put so much of his life into. He was unsure that he would ever be able to work in the United States again. Perhaps more important, he was being so disparaged in the press and felt so destroyed personally that he thought he might lose his family, too. Those were truly dark days.

"Italy was a lifeline," though, he said. It was there that he focused on how he might make the living-donor liver operation even safer, by learning a technique to take less liver from the donor (using the smaller left lobe, rather than the larger right lobe, which had traditionally been transplanted). Will another donor death occur? Absolutely. This is just how statistics work. But shifting the risk he can control away from the donor has been therapeutic for Miller. He has regained his confidence, his love for surgery. He performs surgery more safely now than he did before, and he does a better job when obtaining donor consent, making it as clear to them as it is to him that they can die.

Miller stresses the importance of being prepared, having protocols and support in place, and making sure the donors and recipients know what they're getting into. Even after his infamous donor loss, this eminent surgeon is as supportive of living donation as he ever was—maybe even more so.

I have thought about this topic every day for the last decade. I have considered how Starzl and Calne and many of the other pioneers were opposed to living donations, and I have thought a lot about whether I would let my wife or my kids donate an organ to me. While I think living donation is one of the most wonderful things someone can do, we need to be extremely careful to avoid coercion. Do the people involved truly understand why they want to donate? What do they imagine the experience will be like? Do they know what the risks are? How will they feel if it doesn't play out the way they expect?

I believe in living donation, and I see the donors as heroes. I celebrate them in the same way I would celebrate someone who ran into a burning building to save a loved one. My role is to help them run into that building as safely as possible. But it is never without risk.

HERE IS AN uplifting story that I always think about when I think about living donation. Torril is one of the most memorable and charismatic patients I have ever taken care of, and to this day I'm proud to have been able to play a small role in her life.

Torril's mother needed a kidney transplant. I initially evaluated Torril's father, who seemed healthy. Then, on his CT scan for his workup, we identified a large retroperitoneal sarcoma, a big cancer in the soft tissue surrounding his kidney. He ultimately had this removed, but it ruled him out for donation. In the end, Torril donated her kidney. She likes to say that the process saved two lives: her mom's because she got the kidney, and her dad's because the donor testing serendipitously revealed his undiagnosed cancer.

This story doesn't have a totally happy ending. About a year after the transplant, Torril's mom developed a blood cancer that was likely related to the immunosuppressive medicines she was taking to prevent rejection. She ultimately died from this cancer, which of course was tragic.

Torril and her husband run an organic farm, and during the year after the transplant Torril's parents lived on the farm and helped run it with them. Having this additional year with her mom, who was healthy and active, was worth it for Torril. In a speech she delivered in memory of her mother, who was cremated upon her death, Torril talked about how in her mother's ashes was a piece of her, too, mixed in for eternity.

Part VI

Today and Beyond

Everyone here has the sense that right now is one of those moments when we are influencing the future.

—STEVE JOBS

| 16 |

Complications

No matter what measures are taken, doctors will sometimes falter, and it isn't reasonable to ask that we achieve perfection. What is reasonable is to ask that we never cease to aim for it.

—ATUL GAWANDE, *COMPLICATIONS: A SURGEON'S NOTES ON AN IMPERFECT SCIENCE*

No book about surgery, particularly one about a field as complex as transplant, would be complete without a chapter on complications. As strong as we surgeons are supposed to act in surgery, we all have to figure out how to deal with complications. Managing them medically is the easy part. The challenge is how to handle them emotionally. Complications sit on your shoulders like a heavy weight, sucking the joy out of your life. And to add to the misery, every day, we have to visit patients who are struggling because of errors we made. Many of them are living in despair, unable to eat, sometimes even with stool pouring out of their bellies.

I will never forget the time one of my mentors in residency, a world-famous thoracic surgeon, had a string of complications during several esophagectomies. He had performed literally thousands of them, but for some reason, three in a row leaked, forcing him to divide esophaguses and bring "blowholes" out in the neck (called spit fistulas). I remember him turning to me and saying, "I am creating fucking monsters."

Some complications are of the kind that sit with you day and night—for example, the pancreatic leak after a kidney transplant I did. To this day I don't know how that happened. Whatever the reason, a couple of days after the surgery, the patient developed fluid in his belly. I wanted to make sure it wasn't urine leaking from the new kidney, so I tapped it and had it tested for creatinine, which would be high if it was urine. It also got sent for the pancreatic enzyme amylase, and the results came back positive—meaning the fluid was leaking from the pancreas. I tried everything to stop the leak, from giving the patient medication, to stenting his pancreas, to resecting portions of it, but in the end, nothing worked. I saw this patient every day for months. I saw him when he seemed to be doing better, and I continued to see him when he was clearly getting worse. He got frustrated with me by the end, and I got frustrated with him. I can distinctly remember hoping, when he was really sick, that he would just die. It does not feel good to admit that. And in the end, he did. (Don't get me wrong. I feel awful that it happened, and I feel responsible, even though I don't know why it happened and can't imagine how I caused it. I was just powerless to fix him, and death provided relief for both of us.)

Some complications are straight-up mental errors. Gary was a retired high school teacher (of physics and chemistry) with a wife and two daughters. He was generally a healthy guy, didn't drink or smoke, and had led a good life. But about twenty years before I met him, he had his gall bladder removed, and after the surgery, he was told there was something wrong with his liver. He underwent a workup and was diagnosed with alpha 1-antitrypsin deficiency. Alpha 1-antitrypsin, a protein made in the liver, is secreted into the bloodstream, where it goes to the lungs to protect them from damage. In patients with this genetic disorder, the protein is abnormal and gets stuck in the liver, leading to cellular damage, ultimately causing end-stage liver disease. Many of these patients also develop lung disease from lack of this protective protein (since it is stuck in the liver), most commonly emphysema.

Gary developed cirrhosis of the liver, and eventually his liver stopped functioning almost completely. By the time I met him, he was extremely ill: His kidneys had failed secondary to his liver, and he was on dialysis. His belly was filling with fluid, requiring frequent taps. He was as yellow as a banana, and he was confused. His advanced illness had moved him to the top of our transplant list, and a liver became available for him just in time.

The surgery started at around 5:00 a.m. His hepatectomy was straightforward—he had a shrunken liver with five liters of ascites. The donor liver was of good quality, but it did have some variant anatomy that needed to be reconstructed on the back table. There were a few other minor challenges in the OR, but ultimately, the liver perfused well and worked right away.

We finished up, and I spoke with Gary's wife, Doris. I thought things had gone well. Gary made slow improvement, and after a few days he was out of the ICU and back up to the floor. His kidney function started to recover slowly. He was so ill at the time of the transplant that I knew he would have a long hospital stay.

Then, about a week later, his bilirubin levels started to go back up, and he started to look yellow again. I ordered an ultrasound: the flow in the vessels looked normal. A day later the bilirubin was still going up, and I got a biopsy. No rejection. Finally, the next day, I decided to get an ERCP (the same procedure Nate underwent). I hate having this done too soon after a transplant, since it involves cannulating the donor bile duct that I had just sewn to the recipient duct. I am always nervous that this might disrupt my anastomosis, but I was concerned Gary had a narrowing (stricture) where I had sewn the two ducts together; and fortunately, our guys are good at it, and if there was a narrowing, they could probably thread a stent across it. When they performed the ERCP on Gary, they saw some sort of friable mass blocking the bile duct. They did a biopsy, but couldn't get a stent in.

Damn. I knew I'd flushed the duct before sewing it together, and it flushed fine. Still, Gary's bili levels kept going up, and a couple of days later I had them try another ERCP. This time, they maneuvered past this mass and got a stent in. All good. The biopsy showed benign hyperplastic tissue. I thought we were out of the woods.

Then, a couple of days later, Gary's blood pressure dropped into the seventies, he was confused, and his bowel movements were black—all telltale signs of an upper-GI bleed. We moved him back down to the ICU, got a breathing tube in him, and got the GI guys to send another scope down there. As expected, the benign mass was bleeding impressively. They injected it with epinephrine (which causes the vessels in it to constrict and stop bleeding) and put in some clips. But they weren't confident.

Gary did okay overnight, but then bled again, aggressively, the next day. His pressure was down again, too. They again went in with the scope, this time trying to burn the bleeding mass of tissue. They were not happy; neither was I. Confident he would keep bleeding, I told his wife, Doris, we would need to go back to the OR. She was nervous; so was I.

I took Gary back to the OR later that evening. We entered the abdomen, and the liver looked beautiful. I mobilized the duodenum and, once everything was exposed, called in my colleague Cliff, who operates on the duodenum. Duodenal leaks are rather deadly, given duodenums' corrosive contents (including bile and pancreatic juices) and the difficulty in closing these leaks. Cliff stood over my shoulder as I opened the duodenum and found the clot. It wasn't actually bleeding at this point, but it looked like it would again. And the mass was clearly nothing to be worried about in terms of cancer. I oversewed it carefully, and then closed the duodenum. Cliff gave me his blessing and left.

Now Phil, my fellow, and I turned our attention to the bile duct—and the stress level in the room immediately went down; this was an operation I had done many times. We divided the bile duct from the liver, just below the level of our old anastomosis and just above where the duct courses into the duodenum. We removed the stent that our friends in GI had placed, and handed it off the field. I asked for a suture to oversew the duct, a 2–0 silk. Phil held the end of the duct up so I could place my tie.

Suddenly, we noticed a rush of blood up closer to the liver. Had we disrupted something? We sucked out the blood and saw that the bleeder was just some small, inconsequential vein that had decided it wanted to be heard. We asked for a stitch, and all was well again.

Now we turned our attention to the next part of the operation: getting the bowel ready to be brought up to the duct. We repositioned our retractors lower and prepared the small bowel we would use to reroute the bile duct. Once this was ready, we replaced our retractors, and got the bile duct exposed. We made a hole in the jejunum near the end of our Roux limb and, using interrupted sutures, sewed the duct to the jejunum. It looked perfect at the end. We irrigated with a liter of warm saline and started closing the abdomen.

We finished closing at about two o'clock in the morning. I spoke to Doris, tucked Gary in in the ICU, and went home with the high you get when a complex operation goes well. I figured Gary was out of the woods.

The next day, I was sitting in my office, going over the case in my brain the way a professional golfer goes through each shot in a round of golf, when suddenly I said out loud, "I never tied the other end of the duct shut!"

I quickly called Phil, but he was out on a plane, procuring organs. I left him a message, asking if we'd forgotten that one tie. The second he landed, he called me and said, "Holy shit! We didn't do it! You gotta take him back!"

I walked down to the ICU. Doris beamed at me from the chair next to the bed. In the bed, Gary looked tired but alive. He was so happy he had the breathing tube out, and he started to thank me, but I quickly stopped him and said, "I forgot one step of the operation. I need to take you back."

He asked, "Am I a victim of a medical error?"

Yes, I suppose you are.

Malcolm Gladwell once wrote an essay for *The New Yorker* titled "The Physical Genius." In it, he writes about Wayne Gretzky, Yo-Yo-Ma . . . and master neurosurgeon Charlie Wilson: "Charlie Wilson talks about going running in the morning and reviewing each of the day's operations in his head—visualizing the entire procedure and each potential outcome in advance. 'It was a virtual rehearsal,' he says, 'so when I was actually doing the operation, it was as if I were doing it for the second time.'"

I, too, often sit in my office or at the scrub sink and "perform" the operation in my brain before doing it in real life. I go over the moves, what things are going to look like, what problems we might get into. Before I do a laparoscopy, I look carefully at the CT scans, picturing the anatomy in 3-D, seeing what all the structures will look like and how they will be related. (This gets easier the more you do it.) I can't fall asleep after a big case because I can't stop my head from reviewing the steps of the operation, as if a tape were playing back the moves. I can actually register what I did well and what I might have done better, which helps me think about the next case. It is probably the same for any field that combines technical skills with mental preparation.

Oh, I did take Gary back to the OR that day. Things were pretty stuck already, one day after the other operation. I dug down and got behind my Roux and found the open end of his duct. Would he have been okay if I had left it open? I'm not sure. I oversewed it and closed him back up. And to this day, almost six years after his transplant, he is doing great.

ONE OF THE things that helps all of us in the transplant field deal with complications is the "M and M," or "morbidity and mortality" conference, a weekly meeting where cases from the previous week are discussed. Virtually every department of surgery has M and M, and all resident training programs are mandated to hold these conferences. They are held primarily for the purposes of learning, teaching, and improving quality of patient care, but I think most of us find them cathartic, too. An M and M is usually run by a senior surgeon, often the chairman or division chief. A resident or fellow presents the case, and other surgeons chime in with questions, comments, recommendations, and, often, criticism. Usually the attending surgeon in the case will at some point give his impressions of what happened, what went wrong, what might have been done differently. There certainly are cases where I think I did everything right; where, given the information I had, I would have done things the same way again. But there have been many others where I wish I had made different choices, where I shouldn't have operated, I should have called for help, I should have made a different choice about where to put a kidney, how to sew the artery to the liver, how to handle the ureter. Each one of these decisions likely led to some suffering for the patient—perhaps a reoperation, another procedure, more time in the hospital, blood transfusions, loss of an organ, even death. Heavy stuff—but talking about it with other people in the field is incredibly helpful, even when they call you out on your mistakes. We are all trying to be better for the next patient we see.

One piece of advice I've always liked is that every surgeon needs to have a metaphorical "box" into which he places all his complications. He should be able to access that box each time he sees the patients with complications and their families, and when he presents a case at an M and M conference. At the same time, a surgeon should be able to close that box and put it away when he goes home to his family. Those who fail to maintain access to the box become cavalier and lose their compassion. Conversely, those who don't have a box can struggle to keep their sanity, can't stop thinking about all the bad things that may have happened or they may have caused, even after hours. Such surgeons often leave the profession entirely, or never really get started once they finish their training. Others limit themselves to small procedures, and call partners in to help at the slightest turn of a hair. Whatever the right strategy, we surgeons have to find a way to live with complications, to learn from them, to help patients get through whatever we may have caused or at least been a part of, and to move on.

Sometimes in surgery things happen even though you did nothing wrong. And these cases test you like nothing else in our field. I remember one winter, a few years ago. I was doing a kidney transplant on a young woman—a girl, really—with IgA nephropathy, an autoimmune disease characterized by deposits of antibody in the kidney, leading to inflammation and ultimately kidney failure. She was about nineteen years old and otherwise totally healthy. She was receiving a good deceased-donor kidney, and I remember it was a right kidney. I know this because right kidneys have short renal veins and are always a bit trickier than left ones. When they come from deceased donors, they are usually procured with a cuff of vena cava attached to the right vein, which allows you to extend the vein if needed. I decided not to extend it in this case because the recipient was small, and I figured it was unnecessary. I remember noting that the vein was quite thin as it entered the kidney; but then, right veins always are. I flushed some fluid into the vein to distend it and noted that it was watertight. There were no major branches to tie, no little bleeders to fix. Jake, an impressive resident, and I started at around 9:00 p.m. The case went really well—once we sewed it in, the kidney reperfused beautifully and started making urine right away. I had that incredible feeling of satisfaction as we were closing, and even though it was almost midnight when we finished, we took our time and gave her a nice plastics closure, with dissolvable sutures.

Later, as I pulled into my driveway and turned off the car's engine, I reached over for my phone. I usually keep it in the cup holder next to me, but it wasn't there. I fumbled around in my jacket pocket; it wasn't there, either. I looked over at my house, where all the lights were off, and pictured my daughters sleeping soundly inside. It was already around 12:30 a.m., but as I started to climb out of the car, I had second thoughts. What if I were needed and couldn't be contacted?

I started the car back up, drove the five minutes to the hospital, and made my way to the OR locker room on the third floor. As I entered, I could hear the faint sound of my phone ringing inside my locker. I pulled it out and saw that I had ten missed calls from Jake. Oh shit.

As I tapped open the text app, I started hearing my name being announced on the hospital's overhead pager. I was being summoned urgently to the OR. My heart jumped to my throat.

I ran to the back of the locker room, pulling off my street clothes on the way, grabbed some scrubs, and started changing rapidly, all the while calling Jake back. He answered and started yelling, "She's bleeding. I couldn't get hold of you!"

I ran into the OR just as Jake and the team were pouring betadine over her exposed belly, which was so distended that she looked pregnant. I also noted how the parts of her that didn't have betadine on them looked pale, like a cadaver's. I looked up at the monitors and saw that her blood pressure was in the 60s, and her heart rate was 150. Several anesthesia residents and staff were squeezing bags of blood through the large IVs they had jammed in her neck.

I turned to the circulating nurse, who was frantically getting the room set up, and told her to set up a back table and start making some ice, just in case we had to pull the kidney out. That way, I could flush the blood out and cool the kidney down to be retransplanted back into the patient if possible. Then I went out to the scrub sink and took a deep breath. As we were scrubbing, Jake filled me in on what had happened.

Shortly after hitting the recovery room, the patient's pressure had started dropping. She got confused, and her belly became distended. It was obvious she was bleeding internally. I couldn't decide whether to yell at him for going to the OR on his own or hug him for taking the initiative.

I cut Jake off in midsentence and started rehearsing with him the steps we would take once we got back in there. I was picturing the rivers of blood we would see. I told him to make sure he had some eye protection, then we smashed through the OR doors and walked over to the scrub table, where the tech gowned and gloved us. We threw some drapes on the patient, and I told the tech to get a bunch of laparotomy pads ready. Then we tore through the patient's beautiful skin closure and cut the sutures holding her fascia closed. I told anesthesia that her blood pressure was probably about to drop even more, and then we cut through the second layer. Blood shot out and nearly hit us in the face. We rapidly opened the whole incision and packed a bunch of lap pads in to stop the bleeding temporarily. Anesthesia started to make a little progress on her pressure. It was still low, but at least we'd avoided an arrest. I was dying to pull out the laps that were soaking up her blood and figure out what was going on, but I waited until her pressure went back up.

Telling Jake to grab a sucker, I pulled the kidney up a bit and peered in. Torrents of blood were coming from the side of one of the vessels. I could actually hear it: audible bleeding. I could tell it was venous. I reached my finger down and somehow got it right on top of the hole in the vein—just as Lillehei told a young Christiaan Barnard to do more than fifty years before.

As Jake suctioned vigorously, I managed to gently place a couple of Allis clamps on the side of the renal vein without obstructing the entire vessel. The side wall had blown out of this thin, short right vein. I'm still not sure why it happened, but whatever the cause, we had the bleeding controlled. And amazingly, the kidney still looked pink and perfused. It was no longer making urine, but that seemed the least of our problems at the moment.

While anesthesia continued to resuscitate her, I considered my options. After much thought, I decided it would be okay to sew up the side of the vein. I did so. And once we had the sutures in place and the clamps off, I relaxed. The kidney was still pink, and the vein was soft. Everything was okay. As I dropped the kidney back down into her belly, I asked anesthesia how we were doing. They were back down to one resident and one staff, so that was a good sign. They had rapidly given eight units of blood, probably her entire blood volume. I looked around the room. Bloody sponges littered a floor smeared all over with blood. I looked down at my scrub pants; I could see and feel the sticky blood against my legs under my gown.

Over the next few days, we watched this young woman like hawks. She had come out of the OR with a breathing tube and was brought to the ICU. I went and talked to her family, who had been summoned by the nursing staff back to the OR waiting room after having been told everything was fine the first time around. They were concerned but understanding. Her kidney was a bit slow kicking in, but ultimately it came back to life. After the patient woke up, she was pissed that I had put staples in—I hadn't thought it appropriate to spend time on a nice closure after the second surgery. Her recovery was tough, quite a bit more than she was expecting, but when I saw her in my clinic about six weeks later, her color was better than I had ever seen it. She was back to normal—a radiant, beautiful young woman who now had a working kidney.

She asked me how close she'd come to dying. I told her she was about as close as she could be without being dead. She thanked me, and even gave me a hug. She told me that she didn't blame me for the complication. She felt great and was excited to live without dialysis. She even pulled her shirt up to reveal her staples and said she was proud of her battle scar.

Did I do anything wrong there? Probably. Have I made any changes since that case? Well, I probably now make an extra effort to slow down on the back table, take my time to make sure everything is perfect before I start the implant. I also may be quicker to extend the vein, as this allows me to pull up less on the kidney, sew onto the thicker vein as it gets farther from the kidney, and see better.

One other thing: I never, ever forget my phone. Thanks, Jake, for saving my patient's life. If things had played out just a little differently, she might have died. I would have felt like shit, and she would have missed out on an entire life that she now gets to enjoy—hopefully forever off dialysis.

17

Xenotransplantation

From One Species to Another

Xenotransplantation is the future of transplantation, and always will be.

—NORM SHUMWAY

Xenotransplantation is just around the corner, but it may be a very long corner.

—SIR ROY CALNE, 1995

New Orleans, January 1964

When thirty-eight-year-old surgeon Keith Reemtsma first met Edith Parker, he must have known she was going to die. A twenty-three-year-old black schoolteacher from a small town in Louisiana, Parker was making almost no urine by that point, having developed renal failure from primary kidney disease. She would spend more than two months in the hospital, being maintained on peritoneal dialysis, a new and temporary technique at the time. This was not a long-term solution for her. Unfortunately, she had no potential living donors. Reemtsma did have one option for her, but it was pretty risky: he would offer her both kidneys from a chimpanzee.

The charismatic Reemtsma had already gained a reputation as a risk taker. He had gone to medical school at the University of Pennsylvania and completed his residency at Columbia. After his internship, he joined the military, serving with the navy and marines in the Korean War. In fact, many who knew him personally continue to think he was the model for the Hawkeye Pierce character in *M*A*S*H*. He showed up in Korea with a footlocker full of scotch and was known for his wisecracking and for mixing top-notch martinis. In 1957, he arrived at Tulane, by this point fully trained, and motivated to make his mark on the world.

Not much is written about how Reemtsma got involved in transplantation prior to his experiences with xenotransplant. But on October 8, 1963, in a rarely reported case, he placed both kidneys from a rhesus monkey into a thirty-two-year-old-woman in renal failure. Although the surgery went well, the woman's body rejected these kidneys, and they were removed ten days later, two days before she died from untreatable kidney failure. Still, Reemtsma wasn't deterred. He knew that chimpanzees share a high percentage of genes with humans (some studies have estimated 96 to 99 percent of DNA), higher than with rhesus monkeys (likely closer to the low 90s). So, when a forty-three-year-old dockworker named Jefferson Davis was diagnosed with end-stage renal disease secondary to hypertension, and peritoneal dialysis was begun, Reemtsma knew the clock had started. Davis was suffering from heart failure secondary to massive fluid overload that couldn't be controlled with the short-term dialysis and fluid restriction. Reemtsma had many conversations with

Davis over several weeks, in which they discussed the possibility of a transplant from a chimpanzee. Davis agreed to proceed. He had no other options.

So, on the morning of November 5, 1963, Reemtsma went over to Charity Hospital, next door to Tulane, "and shaved a chimpanzee that had been discarded, because of irascibility, by a circus." He was large enough and shared a compatible blood type with Davis. Reemtsma then brought the ape over to Tulane and proceeded, under anesthesia, to remove both his kidneys en bloc, meaning both were still connected to the aorta and vena cava. He then proceeded to sew these into Davis, the lower end of the aorta and vena cava onto the external iliac artery and vein, and then the ureters separately directly into the bladder (using the same technique Alexis Carrel developed fifty years before, and the same technique we still use when transplanting pediatric en bloc kidneys from a baby into an adult). Davis was treated with azathioprine, steroids, and radiation, which was all the immunosuppression available at the time. He had a rejection crisis four days after transplant, but it responded to treatment with radiation and more steroids. He recovered kidney function and was discharged on December 18, a month and a half after transplant. Sadly, he was readmitted just two days later with pneumonia, which ultimately killed him sixty-three days after his transplant; at the time of his death, his kidneys were functioning fine, with no rejection.

Just a week after Davis's death, Reemtsma performed an identical transplant operation on Edith Parker, again giving her two kidneys from a single chimp. The kidneys functioned right away, making seven liters of urine in the first day. Her blood pressure quickly normalized, as did her labs. Her leg swelling dissipated. She suffered a rejection crisis twenty-three days after her initial transplant, but it responded to treatment. Parker was ultimately discharged home, and returned to her life, resuming her job as a teacher. At six and a half months she was determined to have normal kidney function. Sadly, a full nine months after her transplant, she died suddenly, possibly due to an electrolyte imbalance. An autopsy revealed no evidence of rejection or any other problem with her new kidneys.

Over a two-year period, Reemtsma and his team performed a total of thirteen chimpanzee kidney transplants, with survival generally recorded between nine and sixty days. By 1965, as chronic dialysis was now available, and outcomes with cadaveric transplantation (now called "deceased-donor transplantation") were improving, Reemtsma discontinued xenotransplantation in humans but continued to research it in the laboratory.

Reemtsma's success inspired others to toss their hats into the ring. Thomas Starzl quickly performed six kidney transplants from baboons, with a survival of nineteen to sixty days. Most of his patients died of infection, possibly due to the increased immunosuppression required to maintain these more genetically disparate organs in humans. Numerous other single attempts were performed around the world, perhaps the most famous being the case of baby Fae, who received a baboon heart as an infant at the Loma Linda University Medical Center, in California, in 1984; she survived twenty-one days, and died of rejection. The original plan had been to use this heart as a bridge until a compatible human heart could be located, but an appropriate heart couldn't be found in time.

By the 1990s, researchers had called for a moratorium on clinical xenotransplantation until there could be more discussion and understanding of the risks of transmission of infectious agents, the complexities of informed consent (which seems a bit odd, given that it is hard to imagine a primate or pig consenting), and animal welfare issues. While the FDA issued no official moratorium, it made clear that it would have to approve any future trial of xenotransplantation in humans.

TODAY, NO ONE in the xeno world is thinking about primates as a source for organs for transplant. There are numerous reasons for this. For one, chimpanzees are an endangered and protected species. Also, the limitations of primates as organ donors are several, including the small size of many of their organs (which was what necessitated both kidneys being used for a single recipient) and the fact that these animals are difficult to breed, have only one offspring at a time, are expensive to procure and care for, and are too much like humans for us to get our heads around trying to raise a colony of them just to provide us with organs. Another limitation, one that has gotten a lot of press (although it may be overstated), is that exposure to these animals so genetically close to humans may introduce xenoviruses or other infections that could pose a public health hazard. Perhaps our immune systems would not be ready to fight off the endogenous viruses or other infections found in primates.

Virtually everyone in transplantation now believes that pigs (specifically, miniature pigs) will serve as organ donors if xenotransplantation becomes a reality. They are easy to breed (each litter might have four to eight piglets), their size is appropriate for their consideration as organ donors (twenty to seventy kilograms), there is a fair amount of genetic homology to humans (although nothing in the range of primates), they are cheap, and maybe most important, it is already socially acceptable to breed them for consumption.

Perhaps the biggest challenge in considering pigs as donors for humans is the presence in them of the alpha-gal epitope, a protein on the cells of nonprimate mammals. This protein is absent in humans, apes, and Old World monkeys (due to evolution), and in fact, we have natural antibodies to it, which leads to rapid rejection of grafts from these animals. In 2002, researchers cloned the first mini-swine lacking alpha-gal, a major step forward in the effort to make xenotransplantation a clinical reality. However, it wasn't a magic bullet. Transplants into primates using organs from these pigs survived longer than ever before, but graft function was still measured in days or months rather than years, and extremely severe immunosuppressive regimens were required.

Alpha-gal was not the only barrier to success—it seemed that the immune response to pig organs even without this protein present was still stronger than what we saw in allotransplants. With this barrier, and the discovery of a retrovirus omnipresent in pigs (sadly named PERV, for porcine endogenous retrovirus), excitement about the future of xenotransplant, along with industry funding, essentially dried up.

This all changed with the discovery of CRISPR/Cas9, the gene-editing system that can remove genes with reliability from the embryo of an animal (or even insert new ones), making it possible to generate modified animals ready for experimentation in a matter of months. Since this discovery, numerous advances have been made over the last few years. George Church's group at Harvard successfully generated pigs with all copies of PERV inactivated, a huge accomplishment and one that allayed fears over the possibility of unleashing xenoviruses on humankind, a potential major barrier to FDA approval of trials in xenotransplantation. Multiple companies have been formed that are rapidly modifying the genes of pigs to make their organs look more human, knocking out proteins in an effort to minimize the chances of rejection after transplantation.

One pharmaceutical company, United Therapeutics, has pumped more than a hundred million dollars into similar efforts, partnering with academic leaders at universities, including Alabama and Maryland. The company also has plans to break ground later this year on a massive farm with the capacity to produce a thousand pig organs per year. The farm even has spots for helicopter pads, so organs can be flown out as needed.

So, is this going to happen? Maybe. There has been a conglomeration of scientific superstars and clinical leaders in transplantation/xenotransplantation at a few centers, and industrial support has been pouring in. While the survival rate for life-sustaining organs such as kidneys from pigs into primates has been on the order of a year or more, intense immunosuppression is still required. Despite all the barriers, though, we've come a long way.

As I look at the investigators involved in these efforts, I can't help but be reminded of the pioneers who made transplant a reality despite all the resistance: Starzl, Murray, Shumway, Barnard, Hume, Moore, and the others. I see all the same qualities: drive, focus, confidence that things will work, the courage to try. I predict we will see trials in the next few years. I also predict that the outcomes will be okay but not great. With the introduction of every new procedure, drug, or technique, there is always a learning curve, the black years when you have to go from the occasional success to a true and realistic option for patients. Will these new pioneers be able to hang in there, stay positive, and keep up their courage during the struggle? We shall see.

18

So, You Want to Be a Transplant Surgeon?

He dressed and found himself thinking about the operation again. Should I have tacked the sigmoid colon to the abdominal wall to prevent it twisting again? Didn't I see Stone do this? Colopexy, I think he called it. Had Stone spoken to me about the danger of a colopexy and warned against it, or had he recommended it? I hope we took out all the sponges. Should have counted once more. I should've taken one more look. Checked for bleeders while I was at it. He recalled Stone saying, When the abdomen is open you control it. But once you close it, it controls you. "I understand just what you mean, Thomas," Ghosh said, as he walked out of the theater.

—ABRAHAM VERGHESE, *CUTTING FOR STONE*

Finally, there is something about the practice of surgery that has meant the most to me—more than the intellectual challenge of solving a puzzle, more than the rewards of trying to help others, and more than the gratitude of those you have tried to help. In our patients, we witness human nature in the raw—fear, despair, courage, understanding, hope, resignation, heroism. Our patients teach us about life. In particular, they teach us how to deal with adversity.

—JOE MURRAY, *SURGERY OF THE SOUL*

Back when I was a third-year medical student, I started thinking about a career in surgery. I can't say I had any earth-shattering revelation regarding choosing this specialty. I'd just enjoyed my time on the rotations. I liked the intensity, the idea that you had to train really hard, but that, eventually, you would have a really special skill that would allow you to open people up and fix things. I liked that surgery deals primarily with problems that can be solved, as opposed to managing long-term conditions that never really go away. And I thought surgery was ballsy. I'd always felt that I was calm under pressure, and I was really curious whether I could be that way as a surgeon. And I'd always loved Hawkeye Pierce.

There is a book that is famous among medical students, titled *So You Want to Be a Surgeon*. It's filled with helpful information on how to apply for residency, what steps to take to make your application stronger, which programs to consider. It also lists the character traits that fit a surgical personality: you should like "working on a team"; "embrace responsibility and the opportunity to make a positive impact"; "share the excitement of a surgical team anticipating a great case"; "enjoy watching your patients improve daily after major injuries or surgical procedures."

At the time I first started thinking about becoming a surgeon, I already knew about the excitement of working on a team, the rapid pace at which a surgical service moves, and what it was like to anticipate a big operation. What I didn't yet understand was the level of responsibility that comes with making so many decisions every day that can have a major impact on people's lives. I also didn't realize how much time I would spend worrying about those decisions, how guilty I would feel about mistakes I made, or how stressed out I would be watching patients struggle after operations even when everything went well. I assumed that by the time I was an attending surgeon, I would have a massive bank of information and experience that would guide me through anything. But by the time I got to my chief year (the last year of residency), I realized that you never get that blast of enlightenment, that moment when suddenly everything becomes clear. You just become more comfortable putting together whatever information you have, which is always too little, and then making decisions that are based more on your gut than anything else. Now, after more than a decade on staff, and two decades after medical school, I feel the same way. I have made thousands—no, *millions* of decisions about patients, some big, some small, some right, some not, and almost all of which had some consequence. Many—no, the *majority* of those decisions were right, but so many were wrong. Most of my patients have done well, and yet I can vividly remember almost every one who didn't. I can remember what they looked like when they were suffering or dying, the desperate sadness of their families, who felt helpless to make them better.

Those of us in surgery develop a coping mechanism for dealing with bad outcomes, including blaming our patients or those around us, drinking large quantities of alcohol, or not thinking much about bad outcomes at all. Still, we get a lot of support from our colleagues, and some of us find comfort in presenting our data to the local or national community. In the field of liver transplantation, where patients have become much sicker, and where severe complications and postoperative deaths have become more common, I realize that you can get used to bad outcomes. When you go see a liver recipient before a transplant, and he is in the ICU, with a breathing tube down his throat and an IV drip keeping his blood pressure compatible with life, all you can think is, *I'll give it my best shot. He will surely die without a transplant.*

Of course I want all my patients to do well, and of course I feel empathy and sadness, not to mention responsibility, when talking to the families after a patient dies or has a major complication. At the same time, though, after you walk away from the surgery and go back to the office, or at least by the next morning, when the next case starts, you force yourself to move on. You have to. Nevertheless, each bad outcome, each death, each time you tell family members that their loved one is gone, it takes a little more out of you—and makes it just a little bit harder to put away that box of bad outcomes when you go home at night.

I often think about what the pioneers went through in the early days of transplant. They each had bad outcomes almost continually over decades, with the occasional "success" defined by a patient who stayed alive for a year. There was no guarantee transplant would ever work, many of their colleagues in their hospitals and throughout the medical community thought they were crazy, and there was legitimate concern that they could end up in jail. So how did they persist? What type of personality would be appropriate to take on this kind of challenge? Do people like these trailblazers still exist in surgery? And could similar trailblazing happen now?

A lot of different personalities are represented in the pioneers of transplant. Joe Murray was analytical and religious, a believer. After each failure, he'd review the data, figure out what he could do differently, and move on. Despite bad outcome after bad outcome, he never questioned that his team would one day succeed.

David Hume was an energetic whirling dervish who exuded excitement and who, like Murray, never questioned his chances for ultimate success. He inspired those around him, rarely slept, and was always trying something new.

Roy Calne seems to me the least affected of the bunch. He enjoyed those early days of experimentation immensely, and he has a lightness, a sense of happiness almost, when he recounts those experiences. Calne is the right mix of surgeon and scientist, and he has liked both pursuits.

Norman Shumway, like Calne, inherently loved surgery, loved working with his trainees, and had general disdain for publicity and fame. He was happiest when in the OR, and would often tell residents during operations, "Isn't this fun? Isn't this easy? What could be better? Nothing could be better."

As for Christiaan Barnard, he was much less of a natural surgeon, and much less comfortable in the OR. This is likely why he did so few transplants once he achieved the fame and fortune he was seeking. But he certainly was driven. Anyone who completes a surgical residency, does lab work, completes a thesis and a PhD, and learns two languages in two years is driven.

Walt Lillehei was able to deal with bad outcomes just by accepting them and moving on—patients die; such is life. Oh, and he also found that drinking numerous martinis every night helped, too. Perhaps facing his own near mortality at a young age, when he had cancer, caused him to have a closer relationship with death than most.

Tom Starzl was perhaps the most tortured of the bunch. He has been quoted numerous times stating how much he hated surgery, how he couldn't eat or talk before cases, how he always feared he would screw up and kill the patient. Although he was the first to master what might be one of the hardest operations in the world, he was never comfortable in the OR, and he made those around him pay for it. His surgical personality (that is, how he behaved in the OR) was legendary for its harshness. One characteristic that may have served him well but also tortured him beyond belief was his insanely accurate memory. Once, on an airplane procuring organs (in bad weather, with the plane bouncing all over the place), he dictated a research paper. As he spoke into a tape recorder, he would refer to papers he wanted cited, stating, "Cite my seventh paper here . . . cite my twenty-eighth paper here . . . my two-hundredth paper here . . ." By the time the plane landed, he had completed the paper. Starzl also never forgot a patient, a bad outcome, the faces of a grieving family. He had no coping mechanism to deal with bad outcomes.

Despite their diverse personality traits and coping strategies, all the pioneers had one thing in common: courage. In their book on transplant surgeons, *The Courage to Fail*, leading sociologists and bioethicists Renee Fox and Judith Swazey write that it was courage that got these pioneers through the initial period of transplant in the 1950s and '60s, when it was truly a pipe dream. It was courage that sustained them through the dark years of the 1970s, when outcomes were as bad as 20 to 50 percent chance of one-year survival, with many of the patients suffering miserable deaths secondary to infection and overimmunosuppression—a period that didn't end until cyclosporine became a clinical reality in the early 1980s. According to Fox and Swazey, the pioneers who persisted despite the bad outcomes and the ridicule from colleagues and the public could live with this failure, and never gave up in the battle against death.

No doubt courage is at the heart of what drives a surgeon. But is it the courage to *fail*? I would argue that what separates the pioneers from the rest of us is the courage to *succeed*. Despite all the failures, despite the people around them telling them they were crazy, that they were murderers; despite the threats of dismissal, loss of their medical license, and even imprisonment, they never questioned for a second that they should persevere. This confidence, this courage, I believe, was built into their DNA, and they sustained it for years and years.

Some of the pioneers were also addicted to the act of surgery itself, couldn't get enough of it. Shumway was one. "Surgery, not just heart surgery but all kinds of surgery, is so fascinating, and the responsibility is so acute, that it's a terrible addiction," he wrote. "I was just too consumed by it, and loved surgery so much." Lillehei, Denton Cooley, and Roy Calne all spoke of loving the act of operating. Hume and Moore loved everything about being surgical leaders and innovators, whether it was in the operating room, the lab, or the classroom. For these men, life as a surgeon was all they'd ever wanted; it defined them in every way. Indeed, they had little interest in life outside the OR.

Courageous pioneers still exist in the field. Dr. Nancy Ascher, chief of surgery at UCSF (the surgeon who donated a kidney to her sister), spent many years at the University of Minnesota as it was building its excellent liver transplant center in the 1970s and '80s. She is known as a master surgeon who loves to operate. She keeps her OR totally quiet, so she can focus on (and be stimulated by) the operation at hand. Despite being in surgery for more than forty years, she still loves the actual task of operating, and her focus and her addiction have not dwindled in the least. She and her husband, John Roberts, ultimately went to UCSF in 1988 and built their liver program into one of the premier programs in the world. She also is one of the national leaders in living-donor liver transplantation, one of the most demanding disciplines.

I spoke with Dr. Ascher about the burden of being a surgeon, the responsibility that never goes away. Sure, she frets about patients, she told me, but she's never seen that as a negative. It is an honor for her to perform her patients' transplants, to take responsibility for their new organs, their surgeries, and their outcomes.

Allan Kirk, chairman of surgery at Duke University, last year reached the summit in academic medicine: election to the National Academy of Medicine. When Dr. Kirk was at Wisconsin, he was one of the few fellows able to conduct a clinical fellowship while still working in the lab on experiments with primates. One project he spearheaded during his training involved hooking up patients with fulminant liver failure to pig livers in order to filter the blood until the patients could get a human liver for transplantation. He did this a number of times, but one particular case was of a young girl whom he kept hooked up to a pig liver in a bucket for ten days! He never left the girl's bedside, watching her blood flow out her femoral vein, through plastic tubing into the pig portal vein, and back out the cava and into a vein in her neck. She finally got a transplant, which went well, but sadly, she died afterward. He did ultimately have one long-term survivor of this cross-perfusion procedure, a seventeen-year-old girl who went on to graduate college and have a child after her human liver transplant. Nevertheless, the cross-perfusion procedure was ultimately abandoned due to its complicated nature and unclear benefit.

For Dr. Kirk, the practice of surgery is as much about the science and preclinical experimentation as it is about the operations themselves. If Dr. Ascher is addicted to the practice of surgery, Dr. Kirk is addicted to the life of an academic surgeon. He likes to operate, but the surgery alone is not what drives him.

I feel differently. While I like the challenge of operations, I am certainly not addicted to operating. In fact, I'm always happy when a case is canceled, just as I'm happy when any meeting I'm supposed to attend is canceled. And while I like to push myself to be the best I can, and to accomplish academic work on top of my surgical career, I still don't feel driven, or obsessed, the way pioneers must be. I have the courage to fail but maybe not the courage to succeed, the undying belief that I will always succeed—at least not against all odds.

Pioneers in the field of transplantation were and are special, driven *beasts*, as I like to call them. We all owe them a deep debt of gratitude, and I feel a sense of awe when I think about them.

But even if modern pioneers still exist, could the kind of experimentation and bad outcomes of the early pioneers be tolerated today? Roy Calne's answer to this question was resolute: "Impossible." Starzl felt differently. He thought it was already happening in other fields, such as cancer therapy and gene editing. "It happens right now in front of our eyes, and somebody suddenly does something unexpected, and wow, it's all done. Yes, could happen again, has to happen again."

Things are certainly different from how they were. A surgeon can't just throw a chimpanzee heart or a baboon kidney into someone—and maybe that's a good thing. A surgeon can't just take organs from someone who shows up dead in the ER without talking to family members—again, a good thing. Just because someone is really sick, even dying, doesn't justify trying something new on him or her without having some data supporting a chance of success. And again, this is the way it should be.

Franny Moore would have agreed. In reference to the controversial use of the world's first mechanical heart, in 1969, he wrote:

Desperate measures like the interim substitution of a machine heart . . . call up for consideration a special ethical question . . . does the presence of a dying patient justify the doctor's taking any conceivable step regardless of its degree of hopelessness? The answer to this question must be negative . . . There is simply no evidence to suggest that it would be helpful. It raises false hopes for the patient and his family, it calls into discredit all of biomedical science, and it gives the impression that physicians and surgeons are adventurers rather than circumspect persons seeking to help the suffering and dying by the use of hopeful measures . . . It is only by work in the laboratory and cautious trial in the living animal that "hopeless desperate measures" can become ones that carry with them some promise of reasonable assistance to the patient.

Moore truly was a man ahead of his time, one who believed in surgery as much as anyone ever has. But he also understood the limitations of aggressive surgery, beyond which only suffering and prolongation of misery can ensue.

I'm with Starzl and Moore on the prospects of progress in our current era. Backed by careful laboratory study and sharing of data at national meetings, our modern pioneers, filled with courage, will keep our field marching forward. It is likely that, in the next few years, we will see trials in xenotransplantation using genetically modified pig kidneys in humans. We have numerous tolerance protocols, where patients receive a short course of immunosuppression and then are taken off immunosuppression completely, that work on a small scale, and more and more ideas are entering the clinical arena every year. In addition, novel immunosuppressants we've introduced into the clinic in the last decade are showing promise for our efforts to develop less toxic strategies to keep organs going.

There are so many things I love about being a surgeon. There is nothing quite like spending countless years mastering a trade, finally reaching a point where you alone can open people up and save them when they are at risk of dying. And of all fields, transplant is the best. Every time I put an organ in and it works, I can hardly believe it. I love the fact that we take something from death and give something of great value to the donors and their families. I am truly honored to be a part of this wonderful gift between two people who will forever be connected. On top of all that, I love immunology, the true science of transplant.

The only downside is the never-ending sense of responsibility, the knowledge that despite all the victories, there are also many failures, some of which happened because of decisions you, the surgeon, made. As Starzl said, it does prevent you from living a "normal" life. I often have the feeling that my job is to fix people with illness so they can go back out and live their lives, do things that I would love to do but don't have time to. I have trouble turning my brain off when I get home, trouble living in the moment. My head is always spinning with thoughts of a patient who is having a problem, of some test or procedure I'm waiting on, of a call from a resident about a blood pressure that is too low or a temperature that is too high. I have read so many quotes by surgeons who say that when they look at a patient with an illness, someone whose life is on hold, they feel racked by uncertainty about the diagnosis, and envious—envious of this patient lying in bed resting. He has no responsibility other than his own illness. He has been freed from his worldly responsibilities. No doubt the experience of a major illness is miserable. It separates you from your friends and family, not to mention causing you pain and other physical suffering. At the same time, it releases you from the demands of your life, all the things you have to do every day that you wish you could blow off.

I don't regret my training, my practice, my choices—in fact, I am grateful for the responsibility and the privilege that come with being a surgeon. I am grateful also to play a small role in carrying the baton that the transplant pioneers have passed on to us. At some point, we will hand that baton off to another generation of driven, courageous transplant surgeons, who will move the field to dimensions we can't even dream of.

At the same time, I feel a sense of relief when my kids say they will plan on doing anything with their lives but what I do. Still, they are young. In the end, they will probably both become transplant surgeons. And I will be very proud.

AS THE PLANE levels out at twenty thousand feet, I look over at Felix, our German-trained procurement surgeon, stretched out comfortably in his seat, headphones on, fast asleep. Wide awake myself, I look out at the night sky, now lit up by the moon. For a second, I wonder what I'm doing here, flying in this little plane over the farmlands of Oshkosh, separated from my family. My two little girls, now sleeping peacefully at home, will wake up in the morning and discover me gone, an experience they're quite familiar with.

I shift my gaze back to Felix, along his extended legs, and down to the cooler he is so unceremoniously using as a footrest. Inside sit a liver, two kidneys, and a pancreas. Somewhere else in this same night sky, under this same moon, two other planes are flying in different directions from ours. Each one has a cooler: one contains a heart; the other, two lungs.

Just two days ago, these organs were working in concert, allowing a forty-two-year-old father to eat, drink, go to work, hold his kids. These organs helped him climb up on his roof to clean out the gutters—but they couldn't stop him from falling off. Now they sit in ice, until they will be filled with blood and returned to the living, ready to sustain a new body, allowing five other people to live, love, be happy, be sad, enjoy their families. These five people don't know one another, don't even live in the same city, but they will forever be joined by the web of transplantation. They will be saved by some guy who will never get to see his gift of life. But maybe, down the road, his wife and children will. And maybe they'll think, *Yes, he was some guy.* I know that's what I think.

I lean back and close my eyes. But I can't sleep.

Acknowledgments

It is hard for me to remember a time when books were not a part of my life. I credit this to a rule my parents instituted when I was a child: we had to read two books a week. As I remember it, my parents would also read the books my two brothers and I had selected, and we'd then have lengthy discussions about them at the dinner table. My parents limited the TV we watched, but for some reason they always allowed us to watch *M*A*S*H* (as long as we had read our weekly two books). To this day, I tell people I'm a "meatball surgeon," and I've always wanted to visit Crabapple Cove, Maine, home to my hero Hawkeye Pierce.

I can still remember the moment the spark that would become this book was ignited in my brain. I was sitting on a boat in Miami, late at night, reading *The Emperor of All Maladies*, by Siddhartha Mukherjee. Everyone else had gone to bed, and I knew I had to get up early in the morning for some family activities, but I was enthralled. As I read through the night, fascinated by Mukherjee's ability to use his patients' stories to recount the history of cancer, to make the story accessible to those of us who know little about the topic, I thought about my own field of transplantation, with its incredible history and its ingenious founding fathers. I thought about how new the field of transplant really is, how in the 1940s and '50s, transplantation was truly a pipe dream; in the '60s, it started to happen; and by the '80s, it had become commonplace. I also realized at that moment that some of those pioneers who'd given up so much to make transplantation a reality were still alive to tell their stories.

As I started to put together my thoughts on how I would tackle this project, another book gave me some clues: *My Age of Anxiety*, by Scott Stossel. While Mukherjee uses his patients' stories to tell the history of cancer, Stossel uses his own experiences and struggles to recount the history of anxiety. It was at this point that I decided I would use my coming-of-age as a surgeon and my patients' stories to illustrate the drive and commitment of transplant's pioneers. My hope is to make the story of transplantation accessible to those outside the field of medicine, in a way Mukherjee and Stossel did before me with their own topics.

There are many people to thank for supporting me in this project. I will start with my brother Ben. I have watched Ben with fascination in his career as a writer, which started when he was in high school. I remain amazed that he has been able to keep going for so many years, in such a competitive field, without developing an ounce of negativity. Ever since I voiced my desire to write this book, Ben has been the most supportive person I can imagine, and without him, I would not have followed through on this project. He helped me obtain an agent and gave me extensive advice on actually writing the manuscript. Now that it is a reality, and I have become a published author, I guess he has to perform an organ transplant. Any organ will do. I'll give him the same advice I give my fellows: Don't screw it up.

Next, my agent, Eric Lupfer. I can only imagine what he was thinking when we first spoke some years ago and he agreed to sign on to this project. I'm almost embarrassed to look back at the first outline I sent him after that phone call. (Just as a taste of what it was like, it was titled "The Legend of Big Daddy"—that's me, by the way.) In many ways, he saw what I was trying to do through this process better than I did. I am absolutely blown away by the role he played in molding this book; his positive, relaxed attitude throughout the process; and his gentle (yet firm) guiding hand. None of this would have happened without him. I will be forever grateful.

To my editor, Gail Winston. I remember so vividly the day we met. She immediately understood what I wanted to do with this project, and she was honest about how complex and challenging it might be. She has taught me so much about writing, but a few lessons in particular come to mind: stay disciplined, give the reader credit that they will remember and understand what I have said before, and establish credibility about who I am and why I can tell this story. Gail has so much respect for the readers she works for! I recently read an article on being an author by Thomas Ricks ("The Secret Life of a Book Manuscript," *The Atlantic*, August 22, 2017). In it, his editor tells him, "The first draft is for the writer. The second draft is for the editor. The last draft is for the reader." I'm sorry I sent Gail that massive first draft! I remember telling her to skim it quickly, to make sure it was going in the right direction. After a few agonizing weeks (during which I was picturing her writing to tell me we should just forget the whole thing), she told me that editors don't just skim manuscripts, no matter how rough they may be! I guess I should have warned her I was a Russian language and literature major in college. Even Dostoyevsky would have been impressed with the length and rambling of that one! Thank you so much, Gail, for not giving up on me. And I do plan to pitch a book called *The Cutting Room Floor.* I'm guessing she will pass on that.

To my partners in transplantation, thank you for teaching me how to perform these challenging operations and take care of our patients. Transplantation is a team sport, and we have the best team in the world! Thanks, too, for supporting me through this project—I know it took some extra work from all of you, and I hope you don't make me pay for it.

To all the people I was able to interview, complain to, and argue with through this process—there are too many to name, but nevertheless, I will name some: Sir Roy Calne, Paul Russell, Leslie Brent, David Hamilton, David Cooper, John Daly, Hans Sollinger, Münci Kalayoğlu, David Sutherland, Nancy Ascher, Allan Kirk, Charlie Miller, Arash Salemi, and of course Thomas Starzl; my friend and mentor Joren Madsen, for critically reading the text; the pioneers in our field who gave up so much to make transplantation a reality; our fellows, whom I am so proud of—every year, I am blown away by how well trained you are and how hard you work to become transplant surgeons; it is because of you that I can do my job every day—and Janet Fox, Elaine Snyder, and Mike Armbrust, who helped me track down patients and make our transplant program work.

To my family. To my incredible parents, Molli and Reuben, who have been my greatest supporters throughout my life. Thank you for always teaching me I can do whatever I want as long as I work hard and learn along the way. To my brothers, Jon and Ben, and their families, for their undying love and humor. And to the four people who make my life worth living every day: Gretch, Sam, Kate, and Phoebe (yes, one of them walks on four legs). Gretch, I'll never forget that day I walked into the lab and saw you standing over that pig, doing a lung transplant. (I know you looked at me and thought my shirtsleeves too long.) Transplantation has brought me so many amazing things, but they all pale in comparison to the life we have been able to build together. I owe all my successes to you; your drive, integrity, and quality inspire me daily. You are the most amazing person I have ever met. I am not worthy. Sam and Kate, you are the two best things in my life. When considering all of my accomplishments, being your father is what I am most proud of. I love that you are both obsessed with reading. I never had to institute the two-books-a-week rule with you two—that would actually have *decreased* your volume! (I hope this acknowledgment embarrasses you; little else gives me so much joy!) And Phoebe: sadly, you are my best friend! You have played a large role in this project. Our long, aimless walks around the neighborhood as I worked through story after story were invaluable to this book. It's sad to think you can't actually read this.

To my lower back—you are perhaps the one thing that did not support me in this endeavor. I had always thought surgery was hard on the body. Little did I know that spending hour upon hour hunched over a computer would be even more physically demanding. It may be time for a standing desk . . . or perhaps a back transplant.

To my patients: You are my inspiration. I learn so much from you every day. Thank you for allowing me to be with you while you struggle with illness. Thank you for opening up to me (and for letting me *open you up*), sharing your most vulnerable fears and secrets. It is truly an honor to take part in your care every day, a privilege that all of us in medicine share. We learn more about strength, grace, honesty, and love from you than we could ever give back.

And most of all, to the donors, living and deceased. You are the reason transplantation exists. You are true heroes. I like to believe the donor gets as much out of the act of transplantation as the recipient. As the saying goes, it is better to give than to receive! Thank you for running into that burning building to save someone. Thank you for taking that leap of faith. Your courage gives hope to us all.

Notes

CHAPTER 2: PUZZLE PEOPLE

21   Lyon, France, June 24, 1894: The primary source for this section on Alexis Carrel is the recent book by David Hamilton, *The First Transplant Surgeon: The Flawed Genius of Nobel Prize Winner Alexis Carrel* (Singapore: World Scientific Publishing Co., 2017).

21   Sante Geronimo Caserio: "Caserio Struggled for Life; the Assassin's Courage Failed Him in the End," *New York Times*, August 16, 1894.

21   "How you are hurting me": "Assassination of President Sadi Carnot," *Otago Witness*, Issue 2105, June 28, 1894.

21   An injury to the portal vein: Georg Heberer and R. J. A. M. Van Dongen, eds., *Vascular Surgery* (Berlin: Springer-Verlag, 2012), 4.

23   embroidery lessons at the home of Madame Leroudier: Described in R. Cusimano, M. Cusimano, and S. Cusimano, "The Genius of Alexis Carrel," *Canadian Medical Association Journal* 131, no. 9 (November 1, 1984): 1142.

23   1902 published a paper: A. Carrel, *La technique opératoire des anastomoses vasculaires et de la transplantation des viscères* (Lyon: Medical, 1902), 99859–62.

24   "the crowd of imbeciles and villains": Joseph T. Durkin, *Hope for Our Time: Alexis Carrel on Man and Society* (New York: Harper and Row, 1965).

25   "Carrel patch": Published in A. Carrel and C. C. Guthrie, "Anastomosis of Blood-vessels by the Patching Method and Transplantation of the Kidney," *JAMA* 47 (1906): 1648–50.

26   "Successful Transplantation": A. Carrel and C. C. Guthrie, "Successful Transplantation of Both Kidneys from a Dog into a Bitch with Removal of Both Normal Kidneys from the Latter," *Science* 23, no. 584 (March 9, 1906): 394–5.

27   "perform a series of similar operations": A. Carrel, "The Surgery of Blood Vessels, etc.," *Johns Hopkins Hospital Bulletin* 190 (January 1907): 18–28.

29   Murphy showed in mice: J. B. Murphy and A. W. M. Ellis, "Experiments in the Role of Lymphoid Tissue in the Resistance to Experimental Tuberculosis in Mice," *Journal of Experimental Medicine* 20 (1914): 397–403.

32   "Alexis Carrel": Sir Roy Calne, email to author, November 21, 2016.

CHAPTER 3: THE SIMPLE BEAUTY OF THE KIDNEY

44   At the time, Willem Kolff: The primary source for this section on Kolff is the book by Paul Heiney, *The Nuts and Bolts of Life* (Gloucestershire, UK: Sutton Publishing Ltd., 2002).

CHAPTER 4: SKIN HARVEST

53   "I cannot give any scientist": P. B. Medawar, *Advice to a Young Scientist* (New York: Harper and Row, 1979).

58   North Oxford, England, The Blitz, 1940: The primary source for this section on Peter Medawar is the autobiography by Peter Medawar, *Memoir of a Thinking Radish* (Oxford: Oxford University Press, 1986).

60   published a more complete report: Publication is R. E. Billingham, L. Brent, and P. B. Medawar, "Actively Acquired Tolerance of Foreign Cells," *Nature* 172 (October 3, 1953): 603–6.

60   "biological force": Alexis Carrel, Nobel Lecture, December 11, 1912.

62   "The Fate of Skin Homografts in Man": Publication is T. Gibson and P. B. Medawar, "The Fate of Skin Homografts in Man," *Journal of Anatomy* 77 (1943): 299–310.

63   "'My dear fellow'": Medawar, *Memoir of a Thinking Radish*, 111.

63   Immunogenetics Laboratory, University of Wisconsin, 1944: The primary source for this section on Ray Owen is the article by James F. Crow, "A Golden Anniversary: Cattle Twins and Immune Tolerance," *Genetics* 144 (November 1996): 855–59.

64   Owen published: Publication is R. D. Owen, "Immunogenetic Consequences of Vascular Anastomoses Between Bovine Twins," *Science* 102 (1945): 400–401.

65   published their findings: Publication is R. E. Billingham, G. H. Lampkin, P. B. Medawar, and H. L. Williams, "Tolerance to Homografts, Twin Diagnosis, and the Freemartin Condition in the Cattle," *Heredity* 6 (1952): 201–12.

65   published these findings: Publication is R. E. Billingham, L. Brent, and P. B. Medawar, "Actively Acquired Tolerance of Foreign Cells," *Nature* 172 (1953): 603–6.

65   "The real significance": Medawar, *Memoir of a Thinking Radish*, 133–34.

CHAPTER 5: KIDNEY BEANS: MAKING KIDNEY TRANSPLANT A REALITY

70   Boston, Massachusetts: The primary sources for this section on the origins of kidney transplant are David Hamilton, *A History of Organ Transplantation* (Pittsburgh, PA: University of Pittsburgh Press, 2012); Nicholas L. Tilney, *Transplant from Myth to Reality* (New Haven, CT: Yale University Press, 2003); Paul Terasaki, ed., *History of Transplantation: Thirty-Five Recollections* (Los Angeles: UCLA Tissue Typing Laboratory, 1991); and three autobiographies: Joseph E. Murray, *Surgery of the Soul* (Boston: Boston Medical Library/Watson Publishing International, 2001); Roy Calne, *The Ultimate Gift* (London: Headline Book Publishing, 1998); and Francis D. Moore, *A Miracle and a Privilege* (Washington, DC: Joseph Henry Press, 1995).

72   "administrative objection": Francis D. Moore, *Transplant: The Give and Take of Tissue Transplantation* (New York: Simon and Schuster, 1964), 40.

74 "buzz saw": Thomas Starzl, "My Thirty-Five-Year View of Organ Transplantation," in Terasaki, ed., *History of Transplantation*, 150.

75 "Despite the discomfort": Terasaki, ed., *History of Transplantation*, 39.

76 "Our doubts and hesitation": Ibid., 63.

76 David Hume's series: Published in D. M. Hume, J. P. Merrill, B. F. Miller, and G. W. Thorn, "Experiences with Renal Homotransplantation in the Human: Report of Nine Cases," *Journal of Clinical Investigation* 34 (February 1, 1955): 327–82.

79 "Five months later": Moore, *A Miracle and a Privilege*, 162.

81 "I felt a first blast of heat": Murray, *Surgery of the Soul*, 6.

81 "When I first saw the young aviator": Ibid., 3.

82 "What we were doing": Ibid., 9.

82 "Charles was my introduction": Ibid., 15.

83 "Like many twins": Ibid., 75.

84 crazy "bunch of fools": Ibid., 63.

84 "I had heard of such things": Ibid., 76.

84 "I think we have to be careful": Ibid., 77–78.

85 "We approached insurance companies": Ibid., 77.

85 "Get out of here": Ibid., 78.

93 "Some have wondered": Ibid., 98.

93 "The lecture theatre was crammed": Calne, *The Ultimate Gift*, 21.

94 he found an article in *Nature*: R. Schwartz and W. Damashek, "Drug-induced Immunologic Tolerance," *Nature* 183 (1959): 1682.

94 He published this major finding: R. Y. Calne, "The Rejection of Renal Homografts: Inhibition in Dogs by 6-Mercaptopurine," *Lancet* 1 (1960): 417.

95 "The high point": Calne, *The Ultimate Gift*, 49.

95 "The whole period": Medawar, *Memoir of a Thinking Radish*, 135.

96 "In just eight years": Murray, *Surgery of the Soul*, 117.

97 By 1963, he had presented and published: Publication is T. E. Starzl, T. L. Marchioro, and W. R. Waddell, "The Reversal of Rejection in Human Renal Homografts with Subsequent Development of Homograft Tolerance," *Surgery, Gynecology, and Obstetrics* 117 (1963): 385.

97 Jean-François Borel and Hartmann Stähelin: J. F. Borel, C. Feurer, C. Magnee, and H. Stähelin, "Effects of the New Anti-lymphocytic Peptide Cyclosporin A in Animals," *Immunology* 32 (June 1977): 1017–25.

98 "somewhat reluctantly agreed": Calne, *The Ultimate Gift*, 116.

98 Nevertheless, five of seven: R. Y. Calne, D. J. G. White, S. Thiru, D. B. Evans, P. McMaster, D. C. Dunn, G. N. Craddock, B. D. Pentlow, and K. Rolles, "Cyclosporin A in Patients Receiving Renal Allografts from Cadaver Donors," *Lancet* 2 (December 23–30, 1978): 1323–27.

98 Calne then performed: R. Y. Calne, K. Rolles, D. J. White, S. Thiru, D. B. Evans, P. McMaster, D. C. Dunn, G. N. Craddock, R. G. Henderson, S. Aziz, and P. Lewis, "Cyclosporin A Initially as the Only Immunosuppressant in 34 Recipients of Cadaveric Organs: 32 Kidneys, 2 Pancreases, and 2 Livers," *Lancet* 2 (November 17, 1979): 1033–36.

CHAPTER 6: OPEN HEART: THE INVENTION OF CARDIOPULMONARY BYPASS

103 Open Heart: The primary sources for this section on the origins of open-heart surgery and the development of cardiac bypass are Harris B. Schumacker Jr., *A Dream of the Heart* (Santa Barbara, CA: Fithian Press, 1999); G. Wayne Miller, *King of Hearts* (New York: Crown Publishers, 2000); and David K. C. Cooper, *Open Heart* (New York: Kaplan Publishing, 2010).

113 "Within 6 minutes": Schumacker Jr., *A Dream of the Heart*, 73.

114 "During that long night": Ibid., 74.

115 "I can recall prowling": Ibid., 124.

115 "I will never forget": Ibid., 126.

120 "I shall never forget the first time": Ibid., 154.

125 blood type AB, which no one in his family shared: Blood type AB is the universal recipient, so if it were just a matter of blood donation, a donor with any blood type would be a match. But in this case, where the blood flows through both people once they are hooked up to the catheters, if the person serving as the "pump" were not blood type AB, his own antibodies would have destroyed the blood cells of the baby as they flowed through the circuit.

125 "I just wanted to do": Miller, *King of Hearts*, 155.

126 "I believe we are approaching": Shumacker Jr., *A Dream of the Heart*, 158.

CHAPTER 7: HEARTS ON FIRE: MAKING HEART TRANSPLANT A REALITY

130 Hearts on Fire: The primary sources for this section on heart transplantation are Donald McRae, *Every Second Counts* (New York: G. P. Putnam's Sons, 2006); Christiaan Barnard and Curtis Bill Pepper, *One Life* (Toronto: The Macmillan Company, 1969); and Cooper, *Open Heart*.

136 "[E]ven now I can recall": Barnard and Pepper, *One Life*, 183–84.

137 "When will you sleep?": Ibid., 186.

138 "Look, Chris," Lillehei told him: Ibid., 202–3.

140 "We would stand there": Ibid., 319.

142 "I couldn't leave the patient": Ibid., 329–30.

144 Barnard himself had spent much less time: There is controversy surrounding this. Shumway and Lower were convinced that it was here that Barnard first became acquainted with the idea and techniques of heart transplantation, and then ran back to South Africa to scoop the Americans. Barnard maintains that he was planning heart transplantation for years before this. While it is true that the Americans had spent years doing extensive research in transplantation, on both the operative technique and postoperative care, Barnard himself did do upward of fifty transplants in dogs prior to the first in humans. It is likely that Barnard did pick up some details of the technique from Lower, but unlikely that this short episode played a major role in the success Barnard enjoyed with transplantation.

145 "There is nothing to think": Barnard and Pepper, *One Life*, 259.

146 "If you can't save my daughter": Ibid., 280.

146 "Denise Darvall had entered": Ibid., 286.

147 "So we waited": Ibid., 303.

148 Roughly forty years later: McRae, *Every Second Counts*, 192 and 335. This story remains controversial, and the exact details of what happened that day are open to interpretation. While Marius related these details to McRae, he has reportedly told others that this wasn't the case and that they did wait for the heart to stop before placing Denise Darvall on bypass (personal communication, David Cooper). No one else on the operating team ever confirmed that they actively stopped the heart. Either way, Darvall did meet the criteria for brain death, and starting with Barnard's second transplant, it became routine to stop the heart with potassium.

149 "For a moment": Barnard and Pepper, *One Life*, 314–5.

150 "Eighty-five . . . Eighty": Cooper, *Open Heart*, 335.

151 "Professor, I want": Barnard and Pepper, *One Life*, 392–93.

152 "[T]he prognosis for recovery": McRae, *Every Second Counts*, 277.

154 "The transplant team engaged in": "Virginia Jury Rules That Death Occurs When Brain Dies," *New York Times*, May 27, 1972.

154 "When cerebral function is lost": McRae, *Every Second Counts*, 280.

154 "This simply brings the law": "Virginia Jury Rules That Death Occurs When Brain Dies."

155 "Congratulations on your first": Denton A. Cooley, *100,000 Hearts* (Austin, TX: Dolph Briscoe Center for American History, 2012), 125.

156 "In the eyes of the media": Calne, *The Ultimate Gift*, 128–29.

158 "Although the patient": Wickii T. Vigneswaran, Edward R. Garrity Jr., and John A. Odell, eds., *Lung Transplantation: Principles and Practice* (Boca Raton, FL: CRC Press, 2015), 23.

161 "The appearance of Mary": Sara Wykes, "5 Questions: Bruce Reitz Recalls First Successful Heart-Lung Transplant," Stanford School of Medicine News Center, March 9, 2016.

162 "I said, 'Tom'": Kas Roussy, "Lung Transplant 'Patient 45' Remembered 30 Years Later," CBC News, November 6, 2013.

CHAPTER 8: SYMPATHY FOR THE PANCREAS: CURING DIABETES

163 Sympathy for the Pancreas: The primary sources for this section on type 1 diabetes, including the discovery of insulin and pancreas transplant, are Thea Cooper and Arthur Ainsberg, *Breakthrough: Elizabeth Hughes, the Discovery of Insulin, and the Making of a Medical Miracle* (New York: St. Martin's Press, 2010); Michael Bliss, *Banting: A Biography* (Toronto: University of Toronto Press, 1992); Hamilton, *A History of Organ Transplantation*; and Nicholas L. Tilney, *Transplant: From Myth to Reality* (New Haven, CT: Yale University Press, 2003).

CHAPTER 9: PROMETHEUS REVISITED: LIVER TRANSPLANTS AND THOMAS STARZL

178 Prometheus Revisited: The primary sources for this section on liver transplantation are Thomas E. Starzl, *The Puzzle People* (Pittsburgh, PA: Uni-

versity of Pittsburgh Press, 1992); Calne, *The Ultimate Gift*; Terasaki, ed., *History of Transplantation*; Hamilton, *A History of Organ Transplantation*; and Tilney, *Transplant: From Myth to Reality*.

182 "The truth was worse": Starzl, *The Puzzle People*, 59–60.

186 "The most important consequence": Ibid., 57.

188 "My approach had been": Ibid., 73.

192 "[W]e viewed the principal hurdle": Ibid., 99–100.

201 On April 17, 1968: T. E. Starzl, C. G. Groth, L. Brettschneider, I. Penn, V. A. Fulginiti, J. B. Moon, H. Blanchard, A. J. Martin Jr., and K. A. Porter, "Orthotopic Homotransplantation of the Human Liver," *Annals of Surgery* 168 (1968): 392–415.

201 "A grim conclusion": Starzl, *The Puzzle People*, 170.

201 "The mortality from": Ibid., 165.

203 "After listening": Calne, *The Ultimate Gift*, 96.

204 In 1981, Starzl published: T. E. Starzl, G. B. G. Klintmalm, K. A. Porter, S. Iwatsuki, and G. P. Schroter, "Liver Transplantation with the Use of Cyclosporine A and Prednisone," *New England Journal of Medicine* 305 (July 30, 1981): 266–69.

205 "I was wiped out": Chuck Staresinic, "Only Starzl Dared To," *Pitt Med*, Spring 2006.

206 "maestro" . . . "Watching Dr. Starzl": Quoted in Stephen J. Busalacchi, *White Coat Wisdom* (Madison, WI: Apollo's Voice, 2008), 476.

CHAPTER 10: JASON: THE SECRET IS TO LIVE IN THE PRESENT

215 high MELD score: You might recall from chapter 2 that the "Model for End-Stage Liver Disease" score predicts how sick a patient's liver is, and how likely he or she is to die without a transplant.

219 one of his favorite books: Arthur Herman, *How the Scots Invented the Modern World* (New York: MJF Books, 2001).

CHAPTER 11: LISA AND HERB: SHOULD WE DO LIVER TRANSPLANTS FOR ALCOHOLICS?

224 The six-month rule: R. Osorio, N. Ascher, M. Avery, P. Bacchetti, J. Roberts, and J. Lake, "Predicting Recidivism After Orthotopic Liver Transplantation for Alcoholic Liver Disease," *Hepatology* 20, no. 1 (July 1994): 105–10.

224 To add to the confusion: P. Mathurin, C. Moreno, D. Samuel, J. Dumortier, J. Salleron, F. Durand, H. Castel, A. Duhamel, G. P. Pageaux, V. Leroy, S. Dharancy, A. Louvet, E. Boleslawski, V. Lucidi, T. Gustot, C. Francoz, C. Letoublon, D. Castaing, J. Belghiti, V. Donckier, F. R. Pruvot, and J. C. Duclos-Vallée, "Early Liver Transplantation for Severe Alcoholic Hepatitis," *New England Journal of Medicine* 365 (November 10, 2011): 1790–800.

PART V: THE DONORS

269 "Show me a hero": F. Scott Fitzgerald, *The Notebooks of F. Scott Fitzgerald* (New York: Harcourt Brace Jovanovich, 1978).

CHAPTER 14: AS THEY LAY DYING

271 "Donation gives": Paul I. Terasaki and Jane Schoenberg, eds., *Transplant Success Stories 1993* (Los Angeles, CA: UCLA Tissue Typing Laboratory, 1993), 36–37.

283 "All that remained": Terasaki, ed., *History of Transplantation*, 340.

283 "On June 3, 1963": Terasaki, ed., *History of Transplantation*, 340.

283 "coma depasse": P. Mollaret and M. Goulon, "Le coma dépassé" (preliminary memoir), *Revue Neurologique* 101 (Paris, 1959): 3–15.

284 "We knew that kidneys": Calne, *The Ultimate Gift*, 56.

285 "To throw some fuel": Calixto Machado, "The First Organ Transplant from a Brain-Dead Donor," *Neurology* 64 (2005): 1938–42; taken from J. E. Murray, "Organ Transplantation: The Practical Possibilities," in G. E. W. Wolstenholme and M. O'Connor, eds., *Ethics in Medical Progress: With Special Reference to Transplantation* (Boston: Little, Brown, 1966).

285 "I doubt if": Ibid.

286 "Although Dr. Alexandre's": Ibid.

286 "at the end": Ibid.

286 "Across the country": Murray, *Surgery of the Soul*, 120.

287 He was one of the first: L. Lasagna, F. Mosteller, J. M. Von Felsinger, and H. K. Beecher, "A Study of the Placebo Response," *American Journal of Medicine* 16 (June 1954): 770–79.

287 Beecher became both famous and infamous: H. K. Beecher, "Ethics and Clinical Research," *New England Journal of Medicine* 274 (June 16, 1966): 1354–60.

287 In September 1967: Beecher would later publish the results of this discussion in H. K. Beecher, "Ethical Problems Created by the Hopelessly Unconscious Patient," *New England Journal of Medicine* 278 (June 27, 1968): 1425–30.

288 "As I am sure you are aware": E. F. M. Wijdicks, "The Neurologist and Harvard Criteria for Brain Death," *Neurology* 61 (2003): 970–76.

288 "The subject has been thoroughly": Ibid.

289 The final document: "A Definition of Irreversible Coma: Report of the Ad Hoc Committee of the Harvard Medical School to Examine the Definition of Brain Death." *JAMA* 205 (August 5, 1968): 337–40.

289 "In normal circumstances": Hamilton, *A History of Organ Transplantation*, 347.

290 my patient Wayne: Joshua Mezrich and Joseph Scalea, "As They Lay Dying," *The Atlantic*, April 2015.

CHAPTER 15: HEALTHY DONORS: DO NO HARM

302 Numerous errors: Lydia Polgreen, "Transplant Chief at Mr. Sinai Quits Post in Wake of Inquiry," *New York Times*, Sept 7, 2002.

304 "Italy was a lifeline": Denise Grady, "After Unusual Fatality, Transplant Expert Revives Career," *New York Times*, March 18, 2004.

CHAPTER 16: COMPLICATIONS

314 "Charlie Wilson talks": Malcolm Gladwell, "The Physical Genius," *The New Yorker*, July 25, 1999.

CHAPTER 17: XENOTRANSPLANTATION: FROM ONE SPECIES TO ANOTHER

322 "Xenotransplantation is just": There is some controversy over who should be credited for this very famous quote in my discipline, but I will stick with Roy Calne.

323 He showed up in Korea: Nick Taylor, "Heart to Heart: Can a Chimp Transplant Save Human Life?" *New York Magazine*, July 13, 1987.

324 "and shaved a chimpanzee": Terasaki, ed., *History of Transplantation*, 555.

327 In 2002, researchers cloned: L. Lai, D. Kolber-Simonds, K. W. Park, H. T. Cheong, J. L. Greenstein, G. S. Im, M. Samuel, A. Bonk, A. Rieke, B. N. Day, C. N. Murphy, D. B. Carter, R. J. Hawley, and R. S. Prather, "Production of Alpha-1,3-galactosyltransferase Knockout Pigs by Nuclear Transfer Cloning," *Science* 295 (February 8, 2002): 1089–92.

327 George Church's group: D. Niu, H. J. Wei, L. Lin, H. George, T. Wang, I. H. Lee, H. Y. Zhao, Y. Wang, Y. Kan, E. Shrock, E. Lesha, G. Wang, Y. Luo, Y. Qing, D. Jiao, H. Zhao, X. Zhou, S. Wang, H. Wei, M. Guell, G. M. Church, and L. Yang, "Inactivation of Porcine Endogenous Retrovirus in Pigs Using CRISPR-Cas9," *Science* 357 (September 22, 2017): 1303–7.

CHAPTER 18: SO, YOU WANT TO BE A TRANSPLANT SURGEON?

330 There is a book: Kaj Johansen and David M. Heimbach, "So You Want to Be a Surgeon," 1986, on American College of Surgeons, https://www.facs.org/education/resources/residency-search.

333 "Isn't this fun": McRae, *Every Second Counts*, 67.

334 Once, on an airplane: This story was told to me by Münci Kalayoğlu, who was on the plane with Starzl.

334 In their book: Renee C. Fox and Judith P. Swazey, *The Courage to Fail* (New Brunswick, NJ: Transaction Publishers, 1978).

335 "Surgery, not just": Cooper, *Open Heart*, 348–49.

338 "Desperate measures like": Moore, *Transplant*, 275.

339 As Starzl said: From chapter 10, "The Transplant Surgeon, the Sociologist, and the Historian: A Conversation with Thomas Starzl," in Carla M. Messikomer, Judith P. Swazey, and Allen Glicksman, eds., *Society and Medicine: Essays in Honor of Renee C. Fox* (New Brunswick, NJ: Transaction Publishers, 2003), p. 149.

Selected Bibliography

Barnard, Christiaan, and Curtis Bill Pepper. *One Life*. Toronto: The Macmillan Company, 1969.

Bliss, Michael. *Banting: A Biography*. Toronto: University of Toronto Press, 1992.

Brent, Leslie Baruch. *Sunday's Child? A Memoir*. New Romney, UK: Bank House Books, 2009.

Busalacchi, Stephen J. *White Coat Wisdom*. Madison, WI: Apollo's Voice, 2008.

Calne, Roy. *A Gift of Life*. Lancaster, UK: Medical and Technical Publishing Co. Ltd., 1970.

____. *The Ultimate Gift*. London: Headline Book Publishing, 1998.

Carrel, Alexis. *Man, the Unknown*. New York: Harper and Brothers, 1939.

Cooley, Denton A. *100,000 Hearts*. Austin, TX: Dolph Briscoe Center for American History, 2012.

Cooper, David K. C. *Open Heart*. New York: Kaplan Publishing, 2010.

Cooper, Thea, and Arthur Ainsberg. *Breakthrough*. New York: St. Martin's Press, 2010.

Fox, Renee C., and Judith P. Swazey. *The Courage to Fail*. New Brunswick, NJ: Transaction Publishers, 1978.

____. *Spare Parts: Organ Replacement in American Society*. New York: Oxford University Press, 1992.

Gawande, Atul. *Complications: A Surgeon's Notes on an Imperfect Science*. New York: Picador, 2002.

Hamilton, David. *The First Transplant Surgeon*. Singapore: World Scientific Publishing Co., 2017.

____. *A History of Organ Transplantation*. Pittsburgh, PA: University of Pittsburgh Press, 2012.

Hardy, James Daniel. *The Academic Surgeon: An Autobiography*. Mobile, AL: Magnolia Mansions Press, 2002.

Heiney, Paul. *The Nuts and Bolts of Life*. Stroud, UK: Sutton Publishing Ltd., 2002.

Herman, Arthur. *How the Scots Invented the Modern World*. New York: MJF Books, 2001.

Hollingham, Richard. *Blood and Guts: A History of Surgery*. New York: St. Martin's Press, 2008.

Lederer, Susan E. *Flesh and Blood: Organ Transplantation and Blood Transfusion in Twentieth-Century America*. New York: Oxford University Press, 2008.

Malinin, Theodore I. *Surgery and Life: The Extraordinary Career of Alexis Carrel*. New York: Harcourt Brace Jovanovich, 1979.

McRae, Donald. *Every Second Counts: The Race to Transplant the First Human Heart*. New York: G. P. Putnam's Sons, 2006.

Medawar, Peter B. *Advice to a Young Scientist*. New York: Harper and Row, 1979.

____. *Memoir of a Thinking Radish*. Oxford: Oxford University Press, 1986.

Messikomer, Carla M., Judith P. Swazey, and Allen Glicksman, eds. *Society and Medicine: Essays in Honor of Renee C. Fox*. New Brunswick, NJ: Transaction Publishers, 2003.

Miller, Franklin G., and Robert D. Truog. *Death, Dying, and Organ Transplantation: Reconstructing Medical Ethics at the End of Life*. New York: Oxford University Press, 2011.

Miller, G. Wayne. *King of Hearts*. New York: Crown Publishers, 2000.

Moore, Francis D. *A Miracle and a Privilege*. Washington, DC: Joseph Henry Press, 1995.

____. *Transplant: The Give and Take of Tissue Transplantation*. New York: Simon and Schuster, 1964.

Mukherjee, Siddhartha. *The Emperor of All Maladies*. New York: Scribner, 2010.

Murray, Joseph E. *Surgery of the Soul*. Boston: Boston Medical Library/ Watson Publishing International, 2001.

Najarian, John S. *The Miracle of Transplantation*. Beverly Hills, CA: Phoenix Books Inc., 2009.

Schumacker, Harris B. Jr. *A Dream of the Heart*. Santa Barbara, CA: Fithian Press, 1999.

Starzl, Thomas E. *The Puzzle People*. Pittsburgh, PA: University of Pittsburgh Press, 1992.

Terasaki, Paul, ed. *History of Transplantation: Thirty-Five Recollections*. Los Angeles, CA: UCLA Tissue Typing Laboratory, 1991.

Tilney, Nicholas L. *Transplant: From Myth to Reality*. New Haven, CT: Yale University Press, 2003.

Vigneswaran, Wickii T., Edward R. Garrity, and John A. Odell Jr., eds. *Lung Transplantation: Principles and Practice*. Boca Raton, FL: CRC Press, 2015.

Index

About the Author

A graduate of Princeton University and Cornell University Medical College, JOSHUA MEZRICH, MD, is an associate professor of surgery in the division of multi-organ transplantation at the University of Wisconsin School of Medicine and Public Health.